ART, SPACE
AND THE CITY

Art, Space and the City asks how art and design can contribute to urban futures. It investigates the critical perspectives of cultural geography, urban sociology and critical theory, through analyses of the city, urban space and its gendering, and the monument. A duality emerges between public art which is bound by the aesthetics of the object, and art as a continuous, participatory process of social criticism.

The relation between conventional public art and urban development is one of complicity, whilst emerging practices of art offer resistance to conceptualisations of the city which exclude the interests of its inhabitants. Two roles for art are suggested: as decoration within a re-visioned field of urban design in which the needs of users are central, and as a social process of criticism and engagement, defining the public realm not as public sites but as complex fields of public interest. The tension between these positions is creative.

The book is original in bringing together a range of related but independent perspectives from different disciplines; it goes beyond the restrictions of aesthetic judgement to situate the practice of art for the public realm in terms of social needs and structures of value, and suggests ways in which the perspective of art, as work of the imagination, might also contribute to a critique of the city and to sustainable and convivial urban futures.

Malcolm Miles is Principal Lecturer and Course Director for Design and Public Art at Chelsea College of Art and Design.

ART, SPACE AND THE CITY

Public art and urban futures

•

MALCOLM MILES

London and New York

First published 1997
by Routledge
11 New Fetter Lane, London EC4P 4EE

Simultaneously published in the USA and Canada
by Routledge
29 West 35th Street, New York, NY 10001

© 1997 Malcolm Miles

Typeset in Garamond by Florencetype Ltd, Stoodleigh, Devon

Printed and bound in Great Britain by Redwood Books, Trowbridge, Wiltshire

British Library Cataloguing in Publication Data
A catalogue record for this book is available from the British Library

Library of Congress Cataloging in Publication Data
Miles, Malcolm.
Art, Space and the City/ Malcolm Miles.
p. cm.
Includes bibliographical references and index.
1. Public art. 2. City planning. I. Title.
N8825.M565 1997
711'. 4–dc21 96–48742
CIP

ISBN 0–415–13942–2 (hbk)
0–415–13943–0 (pbk)

CONTENTS

•

List of figures

vi

Acknowledgements

ix

Introduction

1

1 THE CITY

19

2 SPACE, REPRESENTATION AND GENDER

39

3 THE MONUMENT

58

4 THE CONTRADICTIONS OF PUBLIC ART

84

5 ART IN URBAN DEVELOPMENT

104

6 ART AND METROPOLITAN PUBLIC TRANSPORT

132

7 ART IN HEALTH SERVICES

150

8 ART AS A SOCIAL PROCESS

164

9 CONVIVIAL CITIES

188

Notes

209

Further reading

239

Bibliography

245

Index

259

FIGURES

•

1 Jonathon Borofsky, *Hammering Man*, Seattle   6
2 Antony Gormley, one of three cast-iron, double-sided figures on the
   walls of Derry   7
3 Rachel Whiteread, *House*, East London (detail)   9
4 Richard Haas, mural at the Architecture Centre, Boston   10
5 Tess Jaray, paving and street furniture, Centenary Square, Birmingham   11
6 Constantin Brancusi, *Gate of the Kiss*, Tirgu Jiu, Romania   12
7 Joyce Scott, *You Don't Even Know Me*, computer animation, Times
   Square, New York   13
8 Schoolchildren and students from the Kent Institute of Art and
   Design painting a playground mural   14
9 Trafalgar Square   20
10 Villiers Street, London   21
11 Manhattan seen from Battery Park City   22
12 The utopia of new Bucharest   24
13 The Winter Gardens at Battery Park City   26
14 A notice at Battery Park City   35
15 A corporate atrium in Manhattan   40
16 Jim Dine's bronzes referencing Venus   49
17 The Guerrilla Girls, poster commissioned by Public Art Fund,
   New York   52
18 Tourists looking at a bronze Roman Emperor, London   60
19 US Custom House, New York – *Africa*   64
20 US Custom House, New York – *America*   65
21 The feet of the colossus of *Rameses II* on which Shelley's poem
   Ozymandias is based   68
22 Charles Sargeant Jagger's *Artillery Memorial* in Hyde Park, London
   (detail)   69

23 The face of a policeman covered by his helmet, *Cable Street Mural*   72

24 Raymond Mason's *Forward* in Centenary Square, Birmingham   78

25 Kevin Atherton, *Platforms Piece*, Brixton Station, London   79

26 Kevin Atherton, *Platforms Piece* (detail)   79

27 Maya Lin's *Vietnam Veterans Memorial*, Washington, DC   81

28 The *East Coast Memorial*, Battery Park City   82

29 Isamu Noguchi, *Red Cube*, New York   86

30 Richard Serra, *Fulcrum*, Broadgate, London (detail)   87

31 Richard Serra, *Tilted Arc*, New York   90

32 Antony Gormley, *Vision*, Victoria Square, Birmingham   116

33 John Clinch, *The Great Blondinis*, Swindon   120

34 Battery Park City – a waterside landscape and a viewing platform
designed by Mary Miss   121

35 Children at *The Red House*, Sunderland   126

36 *The Red House*, Sunderland   129

37 A relief depicting *Edward II* by children in Swansea   130

38 Valerie Jaudon, *Long Division*; railings designed for the New York
Subway   136

39 Eduardo Paolozzi, mosaics for Tottenham Court Road Underground
station   139

40 Art from the National Gallery in the London Underground   140

41 Art from the National Gallery in the London Underground (detail)   141

42 Tom Otterness, figures sawing a column, for 14th Street Station,
New York   148

43 A poster against the M11 on *House*   149

44 The interior of the new Outpatient Dept at Queen Elizabeth Hospital,
Gateshead, with *Sea Piece* by Mike Davis and Kate Watkinson   156

45 An uplighter, with *Sea Piece*, beyond   157

46 Christine Constant, tile mural of Carlisle at the Cumberland Infirmary   158

47 'We Got It' – a candy bar designed by production workers, Sculpture
in Action, Chicago   168

48 Wood-carving in The Art Studio, Sunderland   170

49 Detail of exhibition space, with work by artists of The Art Studio,
Sunderland, 1996   171

50 Gran Fury, bus shelter installations, New York   175

51 Edgar Heap of Birds, installation in Pioneer Square, Seattle   181

52 Peter Randall-Page, shell form on a footpath in Dorset for Common
Ground   183

53 Somewhere between graffiti and art, on a boarded-up doorway in
Seattle   195

54 Paley Park, New York   196

55 Greenacre Park, New York   197
56 Zoning regulations ensure trees but not sociation – spaces remain too
   regulated and sterile   198
57 Siah Armajani, text in railings at Battery Park City   202
58 Gordon Young, *Fish Pavement*, Hull   204
59 Gordon Young, *Fish Pavement*, Hull   204
60 *House*, shortly before demolition   206

ACKNOWLEDGEMENTS

•

Many people have been generous with their time and kind enough to share with the writer their views on the subject, in particular: Cynthia Abramson, Kevin Atherton, David Butler, Helen Denniston, Peter Dunn, Tom Eccles, Wendy Feuer, Patricia Fuller, Valerie Holman, Mary Jane Jacob, Jeffrey Kastner, Lucy Milton, Peter Randall-Page, David Reason, Miffa Salter and Valerie Swales; two employers – first the University of Portsmouth, then Chelsea College of Art and Design – have been supportive, whilst many of the book's arguments have been honed through discussion with students in both institutions. Friends, too, have helped sustain the writer in moments of self-doubt, in particular Debbie Duffin, Fiona Furness, Anne Harrison, Paul Stickley and Jackie West.

INTRODUCTION

•

THE AIM OF THE BOOK

The aim of this book is to construct bridges between, on one hand, contemporary practices of art and design for urban public spaces, and, on the other, critiques of the city generated in disciplines such as urban sociology and geography, informed by critical theories of society and culture. From this it is possible to speculate on the roles of artists and designers in urban futures.

To date, the specialist practice called 'public art', which includes a diversity of not always compatible approaches to making and siting art outside conventional art spaces – from the exhibition of sculpture outdoors, to community murals, land art, site-specific art, the design of paving and street furniture and performance as art – has grown in isolation from debates on the future of cities, largely untouched by the theoretical perspectives which enliven other disciplines; as a result it is an impoverished field, with little critical writing through which artists and designers can interrogate their practices. Public art is, too, a marginal area within art practice, having little appeal to curators, dealers and critics for whom it lacks the autonomy of modernist and contemporary art and offers few opportunities for the manufacture of reputations, accumulation of profit or demonstration of taste; yet advocacy for public art, supported by the Arts Councils and National Endowment for the Arts, has been successful, and it is commonplace to include provision for art in urban development in the cities of the industrialised states of the west.[1] Since this advocacy has been unquestioning of the intentions of development and its impact on communities, art has perhaps been complicit in the abjection that increasingly follows development and the extension of privatisation and surveillance.

Because public art acts in the public realm, its critique necessarily extends to a series of overlapping issues, such as the diversity of urban publics and cultures, the functions and gendering of public space, the operations of power, and the roles of professionals of the built environment in relation to non-professional urban dwellers. These issues are as relevant to architecture, urban design and urban planning as to public art, even if art is sometimes seen in popular culture as romantic escapism. After three decades of public art as a specialist practice it is time to take stock of what it has become; yet the book seeks to contribute as much to a critique of the construction of cities as to one of public art – the bridges carry ideas in both directions. Bridges are also things we cross *en route*, and the book argues for the application to urban planning and design of the concepts of 'liveability' as developed by urbanists, mainly in the USA, and 'sustainability' as used in discussion of technological aid to 'developing' countries,[2] and sees in the diversity of alternative modes of dwelling summarised by the phrase 'living lightly upon the earth'[3] parts of a model for a new urbanism. Parallel with these approaches are Elizabeth Wilson's call for a re-visioning of urban planning,[4] and Suzi Gablik's proposal for participatory art, which is concerned with social and ecological healing.[5]

The book, then, has two points of departure: public art – the making, management and mediation of art outside its conventional location in museums and galleries – and the convivial city – involving user-centred strategies for urban planning and design. Both are problematic and prone to misconception through non-critical advocacy, but have emerged from different fields and tend to be discussed in isolation. The zoning of knowledge in specialist professional and academic disciplines enables a displacement of accountability, so that decisions which determine city form appear remote from those they affect, or inevitable, like a kind of 'urban blight'; but cities are not like potatoes and have been planned by people regarded as expert. Perhaps it will be in the spaces between disciplines that alternative frameworks will emerge, and perhaps these will be nurtured by the imagination to which the creative arts lay claim, but which is not their exclusive property.

This book is an introduction to a way of looking at art in public spaces from viewpoints outside art; it is not a history or survey of public art and does not advocate public art, for which the claims made tend to be nebulous and the social benefits undemonstrated and perhaps, given the vagueness of the claims, undemonstrable.[6] But, whilst the text questions the effectiveness of public art in contributing to sustainable urban environments, it retains the idea that imagining possible futures (a project no more restricted to professionals than imagination is restricted to artists) is as much part of a democratic society as informal mixing in public spaces, and that such imaginings may produce an urban regeneration in which the social benefits are primary. If this seems credulous, the alternatives – acceptance of terminal urban decline, cynicism, or a flight to an illusory rural paradise – are varieties of despair, and they defeat the aim of criticism to construct a future freed from replication of the past.

For whom is the book written?

The book is intended for both an academic and professional readership, to meet the needs of students, educators and practitioners in fields such as art and craft, architecture and environmental design, urban design, landscape design, urban planning, urban history, urban sociology, cultural geography and cultural policy. Its intention is to be accessible to readers not conversant with the discourses of disciplines other than their own, stating the arguments as far as their complexity allows in ordinary language and offering scope for selective further study through the reading suggested for each chapter. References are listed in the general bibliography, and notes used to add supplementary information or tangential thoughts in a way that does not clutter the main text.

Selected sources

The subject is potentially vast, necessitating a high degree of selectivity in drawing on published sources and documents to articulate perspectives on art, urban space and the city, and in taking cases to illustrate the argument. Not all those whose views are reported agree with each other, and the book seeks to impart something of the intricacy of current debate as well its outlines; whilst it is often possible to summarise a contribution, at times selective citing of texts is used to bring out the diversity of voices which define the field. The selectivity of examples of public art reflects the location – the UK – in which the book is written, and the extent of public art practice in the USA, with which comparisons are most easily made. Many of the illustrations are the writer's original photographs.

The literature of public art is small compared with that of art in museums and galleries. One of the first books on the subject – John Willett's *Art in a City*, of 1967 – remains one of the more questioning compared with works produced two decades later when advocacy for public art reached its peak, such as John Beardsley's *Earthworks and Beyond*, or Deanna Petherbridge's *Art for Architecture*. Somewhere between these positions are *Art Within Reach*, a collection of essays edited by Peter Townsend in 1984, and *The Furnished Landscape*, which accompanied a Crafts Council exhibition of applied art for public spaces in 1992. Critical works on art in public spaces are Nina Felshin's *But is it Art?*; Diane Ghirardo's *Out of Site*; Suzanne Lacy's *Mapping the Terrain*; William Mitchell's *Art and the Public Sphere*; Arlene Raven's *Art in the Public Interest*; and Brian Wallis' *If You Lived Here*. These were all published in the USA from 1991 to 1995, four by the Bay Press in Seattle, and are multi-author works, as are *Mapping the Futures*, edited by a group of academics at Middlesex University (Bird *et al.*, 1993) – which brings together writers from art history, architecture, cultural studies, philosophy, geography and urban sociology – and *Re-Presenting the City*, edited by Anthony King, with contributions from geographers

James Duncan and Neil Smith, planner Saskia Sassen and sociologist Sharon Zukin. Samir al-Khalil's *The Monument* applies critical perspectives to Saddam Husain's monument to a victory never won, and James Lingwood's *House*, with essays by Jon Bird and Doreen Massey, gives a range of perspectives on that controversial temporary work. Mainstream art criticism says little on public art, but Suzi Gablik's *The Reenchantment of Art* is a key contribution.

Amongst theoretical works which have informed the writing of the book are Henri Lefebvre's *The Production of Space*, Michel de Certeau's *The Practice of Everyday Life*, and Michel Foucault's *Madness and Civilisation* and *The Birth of the Clinic*; amongst critiques of the urban situation are Marshall Berman's *All That is Solid Melts into Air*, Mike Davis' *City of Quartz*, Peter Lang's *Mortal City*, and Michael Sorkin's *Variations on a Theme Park*, whilst from urban sociology are *Urban Process and Power* by Peter Ambrose, *Simmel and Since* by David Frisby and *Urban Sociology, Capitalism and Modernity* by Mike Savage and Alan Warde. Elizabeth Wilson, in *The Sphinx and the City*, examines, in part through a personal narrative, the inter-relation of women and cities, giving a critique of both planning and development; from geography are contributions such as David Harvey's *The Urban Experience*, Doreen Massey's *Space, Place and Gender*, David Sibley's *Geographies of Exclusion*, and Edward Soja's *Postmodern Geographies*. Foucault's work on social institutions and the disciplining of publics to produce social order is a pervading influence on Richard Sennett's *Flesh and Stone*; Sennett's previous works – *The Uses of Disorder*, *The Fall of Public Man*, and *The Conscience of the Eye* – relate city life to cultural values, and Indra Kagis McEwen's *Socrates' Ancestor* links city form to the material culture of archaic and classical Greece. The authority of Georg Simmel retains currency, not only through his influence on sociology, but perhaps more within the discourse of modernity alongside Walter Benjamin's Arcades project, recently analysed by Susan Buck-Morss in *The Dialectics of Seeing*. Lewis Mumford's *The City in History* is a modernist reading of the past located in a literature aimed at reforming urban development which also includes Jane Jacob's *The Death and Life of Great American Cities*; an empirical approach to urban design is found in William H. Whyte's *The Social Life of Small Urban Spaces* and *City*, which set out practical means to people-centred urban design, and Roberta Gratz's *The Living City*. Herbert Girardet's *Gaia Atlas of Cities* sees urban form and infrastructure from an ecological viewpoint, and Murray Bookchin explores *Urbanization Without Cities*.

Organisation

The book is arranged in nine chapters grouped in three sections: the first four chapters outline critical and theoretical perspectives on the city, urban space, the monument and public art. The second group of three chapters apply critical perspectives to art in urban development, public transport, and health care. The final section consists of two chapters – on art as a social process, and on the role of decorative and activist art in the construction of convivial cities. The remainder of this introduction problematises, as a prologue for the general reader, the idea of public art, and notes aspects of the contemporary urban context to which it responds, or not.

ART IN PUBLIC SPACES – AN OUTLINE OF THE TERRITORY

Since the late 1960s, works of contemporary art and craft have increasingly been located in city squares and government buildings, corporate plazas, parks and garden festivals, schools, hospitals, railway stations and on the external walls of houses, in a growth of commissioning echoing, in a different visual language and with a broader range of settings, that of statues and memorials in the nineteenth century. Most public art in the UK has been initiated by the public sector – about three times as much as that found in private sector property development (Roberts *et al.*, 1993), and much in the USA and Europe (including the UK) has been commissioned through public bodies. The Arts Councils in the UK promote a Percent for Art policy, through which a given percentage (usually 1 per cent) within the budget for a building scheme is set aside for the commissioning of art or craft works; such policies are currently operated (at levels from 0.5 per cent to 2 per cent) by more than 90 cities and states in the USA.[7] Public art is a major area of state patronage, but the way in which it conveys the state's ideology is seldom overt, concealed in matters of style and the bureaucracies of arts management.

The term 'public art' generally describes works commissioned for sites of open public access; the term 'site-specific' is also used, both for art made for installation in a given site, and art which is the design of the site itself, although in some cases, such as Jonathan Borofsky's *Hammering Man* (Figure 1), or several pieces by Henry Moore, a work is made in a small edition and sited in more than one place.[8] For works which are perceived first as another piece in an artist's *oeuvre* and only second as an adaptation for the site, such as Antony Gormley's iron men on the walls of Derry in 1987 (Figure 2), or perhaps his *Angel* for Gateshead, 'site-general' has been thought a more accurate description.[9]

Most (but not all) sites are urban and the range of works located in them includes sculptural objects such as Alexander Calder's *La Grande Vitesse* at Grand Rapids (the first work to receive a grant through the NEA art in public places programme in 1967), Richard Serra's *Tilted Arc* in Federal Plaza, New York, removed in 1989 after controversy, Rachel Whiteread's intentionally temporary *House* (Figure 3), in East London during the winter of 1993–4, murals, such as Richard Haas' illusionistic treatment of the Architecture Centre in Boston (Figure 4), and integrated designs for paving and street furniture, such as Tess

FIGURE 1   Jonathon Borofsky, *Hammering Man*, Seattle

FIGURE 2 Antony Gormley, one of three cast-iron, double-sided figures on the walls of Derry

Jaray's work in Centenary Square, Birmingham (Figure 5), and Gordon Young's *Fish Pavement* (Figures 58 and 59) – images of a lost industry inserted in the pavements of Hull. Some sculpture, too, is integral to its site, either, like Colin Wilbourne's work in Sunderland, through accountability to a steering group of local people, or like Kevin Atherton's *Platforms Piece* (Figures 25 and 26) – bronze life-casts of three commuters – at Brixton Station in south London, through a local narrative. Scott Burton designed seating for sites in New York and Seattle, uniting the decorative and functional; a one-kilometre axis is formed through the city by Brancusi's work in Tirgu Jiu, Romania, consisting of the *Table of Silence*, *Gate of the Kiss* (Figure 6) and *Endless Column*, collectively a memorial to partisans killed in the Great War, whilst Dany Karavan has designed an architectural axis aligning the satellite town of Cergy-Pontoise with Paris. There are also 'anti-monuments', such as the *Monument against Fascism* by Jochen Gertz and Esther Shalev-Gertz in Harburg, Germany. Jenny Holtzer's texts on electronic message boards and Joyce Scott's computer animation on lightboards in Times Square (Figure 7), billboards by Barbara Kruger, projections onto architecture by Krzysztof Wodiczko, performance pieces such as Suzanne Lacy's *Crystal Quilt*, the range of projects collectively called *Culture in Action*, co-ordinated by Mary Jane Jacob in Chicago in 1993 including *Street-level Video* and the *Chicago Urban Ecology Action Group*, discussed by Jeffrey Kastner under the heading 'art as a verb' (Kastner, 1995), the work of Mierle Laderman Ukeles, for several years an unfunded 'artist in residence' with the sanitation department in New York City, and of Dominique Mazeaud, who repeatedly walked the course of a river clearing it of litter, are critical interventions in the public realm. The 'social sculpture' developed by Joseph Beuys in Germany is also a case of art as process. This sketchy list is given not as a canon, but to show the breadth of the category 'public art', now extending through the interactive potential of electronic technology and the free access of the internet.[10]

Public art might, since definitions are mutable and cumulative, be taken to include the work of artists undertaking residencies in industrial or social settings, and the community arts programmes which began in the late 1960s, or community wall paintings which, in the service of black power, women's rights or movements for national liberation from central America to the north of Ireland, were concerned to make visible the voices of groups who then lacked access to broadcast television; such movements are now seen as roots of what Lacy has termed 'new genre public art' (Lacy, 1995). The work of the Mexican muralists, such as Diego Rivera, has also been taken, with Leger's murals of 1937,[11] as politically committed public art, and has influenced work such as the *Cable Street Mural* in East London. One difference between conventional public art and community arts or new genre public art, is that the community artist, and often the new genre public artist, acts as a catalyst for other people's creativity, political imagination

FIGURE 3   Rachel Whiteread, *House*, East London (detail)

FIGURE 4 Richard Haas, mural at the Architecture Centre, Boston

FIGURE 5 Tess Jaray, paving and street furniture, Centenary Square, Birmingham

FIGURE 6
Constantin
Brancusi, *Gate of
the Kiss*, Tirgu Jiu,
Romania

being perhaps as valued as drawing skill. This is a reaction against the commodi-
tisation of art by its markets and institutions, a rejection of the self-contained
aesthetic of modernism, and reflects a critical realism derived from Marxism, feminism and ecology which implies that artists act for and with others in reclaiming
responsibility for their futures.

All these forms of art practice are located outside the spaces and conventions
of galleries and museums, which is the broadest definition of public art. Definitions
are, in any case, no more interesting than finite, and perhaps a better question
might be whether there are commonalities between public art and popular culture,
since for the majority of people in western countries, television, film and advertising imagery are their major visual intake. Advertising uses subtle codes, for
example promoting brands of cigarettes without naming them, setting a conceptual as well as technical standard against which public art is not always successful
in competing.

Reception

The case for public art claims that it contributes to urban regeneration;[12] but the
case is speculative, and the values of contemporary art are seen as independent of
the problems of city life. The failure of much public art to create a public is

FIGURE 7 Joyce Scott, *You Don't Even Know Me*, computer animation, Times Square, New York (Photo: Oren Slor, Public Art Fund)

linked to its location within the ethos of modernism, in which the interests of artists and publics may be contradictory; modernism imposes constant revolutions of style within a largely static conception of art as constituting an autonomous, aesthetic realm which acts as an alternative to everyday life, and endows the artist with a freedom which Marshall Berman describes as 'a perfectly formed, perfectly sealed tomb' (Berman, 1983: 30). At the same time, Paley Park and Greenacre

FIGURE 8
Schoolchildren and
students from the
Kent Institute of
Art and Design
painting a
playground mural

Park in New York, both 'vest-pocket' parks designed by landscape architects, were observed by William H. Whyte as successful public spaces (Whyte, 1980: 67), whilst neither includes art. Perhaps the challenge to those who advocate the commissioning of public art is to say what it could add in such settings, or perhaps it is to find other ways artists can contribute to liveable cities than by making art objects, for example through intervening as social critics; in either case, the reception of public art is as important as, and linked to, its production.

Works of contemporary art in public spaces are encountered by diverse publics who have, to a large extent, no contact with art in galleries, though they may be adept at reading the codes of mass culture. This raises questions as to the extent to which all art is 'public' and the level of education required if an observer is to understand the work, and whether a site is seen as a physical space, or has another set of dimensions in the public realm. The notion of publicness as site remained largely undisputed until Patricia Phillips' seminal article 'Out of Order: the public art machine' for *Art Forum* in 1988; Phillips attacked the assumption that 'this art derives its "publicness" from where it is located', arguing that the concept 'public' is 'a difficult, mutable and perhaps somewhat atrophied one' and that 'the public dimension is a psychological, rather than a physical or environ- mental, construct' (Phillips, 1988: 93). The reception of public art, then, crosses the gendered boundaries of public and private domains, just as public issues are not bounded by space, and television and electronic media are public in terms of access but consumed in domestic spaces and controlled by corporate interests. The *Domestic Violence Milk Carton* project (1992) by Peggy Diggs, on which Phillips also writes (Phillips, 1995b), exploits this ambiguity.

Is all art public?

The fresco cycles of European churches and cathedrals once had a defined public, but they are not public *art* – many were seen in their time as artisan-work, their value deriving from the repetition of models from tradition. In contrast, art today privileges individualism and subversion of the previous mainstream position. Success in it, according to the late Peter Dormer, 'depends on an understanding of the orthodoxies of the moment and knowing how far these orthodoxies can be manipulated, broken or ignored' (Dormer, 1994: 26). The result is that contemporary art appeals to a specialist public for whom this self-referential development has meaning, but its re-location to public places does not in itself increase access to it more than incidentally, particularly when the outreach activities which have been developed in museums and galleries are lacking. In the nineteenth century, new museums and public collections of art, along with ethnography and natural history, were established as a form of public education appropriate to a reformist society,[13] whilst in the streets and squares of cities the commissioning of statues and memorials proclaimed the manufacture of a national cultural identity. The new museums brought art into the public arena, if reaching a mainly bourgeois public, as, more recently, television, popular films and colour reproduction in the press have brought art to a wide spectrum of viewers. Although the commissioning of art for public spaces may share an ideology with the establishment of museums, both being cases of 'public good' aimed at social stability, the experience of art in public collections and in public spaces differs – in the way art is 'framed' by its location in a cultural institution or in a social setting which is never successfully colonised as an art space; also in that most works collected in national galleries and museums of art were made as commodities for the art market, now re-presented as national heritage,[14] whilst some examples of public art are less given to commodity status, or actively deny it.

Cases of public art for which there is a broad public are scarce, though one is Maya Lin's *Vietnam Veterans Memorial* in Washington, DC (Figure 27), which is constantly visited by people for whom it acts to heal the social wounds of the Vietnam War, whilst it is set physically and conceptually in white American history by being aligned with the Lincoln and Washington monuments on the Mall.[15]

Meaning and mediation

When art is encountered in public spaces, whether a public sculpture such as Henry Moore's *Arch* in Hyde Park, London, or a poster for AIDS awareness, it produces a co-incidence of possible readings of art, city form, and patterns of sociation, which may collide. Whilst dwellers may be in their way 'expert' on their streets, make personal appropriations of elements of the urban fabric in their memories, and feel an affinity with the locality in which they live, the notion that members of diverse publics might be 'expert' on art finds little support in the artworld. It is in art, perhaps more than in other fields,[16] that the role of the intermediary has taken on a dominance, since most public art projects are managed by professional arts managers.

Willett (1967) sets out various intentions of state support for the arts, as histories of art in the service of revolution (France after 1789, Russia after 1917), economic growth (France and England in the nineteenth century), and as a moral force connected to an improvement in the conditions of the poor (England in the nineteenth century),[17] but is guarded on the role of bodies such as the Arts Council, established in 1946 within an emerging British welfare state to support artists and promote the arts in daily life:[18] 'the suspicion is bound to arise that the guardians of art are only trying to find takers for the kind of art that they themselves enjoy' (Willett, 1967: 239). The dominant position of mediators, such as the arts manager, as well as the dealer, curator and critic who make reputations, or the connoisseur who can tell a 'real' Rembrandt, and the links between different kinds of mediators, enforce notions of aesthetic quality as a closed consensus; these notions affect public art as well as art in galleries. Willett draws attention to the divergence, on the evidence of sales of popular prints in Liverpool, of public tastes from a more monolithic, officially supported taste, though he offers as a solution, alongside the education of the eye and inclusion of art in formal education at all levels, risk-taking in commissioning works of 'quality' for public sites, rather than inclusion of popular prints in public collections, which leads us back to the question of by whom 'quality' is determined if not by people such as Willett himself, and whether the most important quality in public art is beauty or empowerment.

There are, then, several problems in the advocacy of public art as a social good: the exclusivity of taste; the lack of specificity of the public(s) for whom it is intended; and the transcendent aesthetic of modernism which separates art from life.[19] Since most criticism within the artworld aestheticises its objects, it is in critical theory and the critical positions of urban sociology and cultural geography that alternative perspectives can be found. In the spaces between disciplines, in which this book is located, each discipline's assumptions can be examined.[20]

A context of urban crises

Discussion of public art takes place in the context of perceived urban decay, in which the utopian aspirations of modernism have been replaced by a post-modern cynicism; perhaps the basis of urban progress, in which urban and economic expansion are mutually identified and facilitated through rational planning, is no longer viable. Architect Richard Plunz writes of Mott Haven in the South Bronx as 'a place that defies even the notion of dystopia' (Plunz, 1995: 29), while in some districts of Los Angeles daily life can barely be sustained and health care is less available than in Mozambique following a post-colonial war. Richard Sennett writes that 'society has come to expect too much order, too much coherence in its communal life' (Sennett, 1996: 181), and describes walking through New York, considering Baudelaire's image of the metropolis and observing the shifting character of neighbourhoods: 'one is immersed in the differences of this most diverse of cities, but precisely because the scenes are disengaged they seem unlikely to offer themselves as significant encounters' (Sennett, [1990] 1992: 128). Perhaps the product of Enlightenment and modernist thought translated into cities of order, free circulation and vistas purged of the unclean, is a compartmentalisation of life and knowledge, producing, in its post-modern parodies, suburban lawns which, according to Mike Davis 'sprout forests of ominous little signs warning 'armed response!' [and] richer neighbourhoods [which] isolate themselves behind walls guarded by gun-toting private police and state-of-the-art electronic surveillance' (Davis, [1990] 1992: 223); architect Peter Lang writes: 'We have gone from fearing the death of the city to fearing the city of death, and I think it is a very traumatic change' (Lang, 1995: 71).[21]

Conclusion

In face of this perception of crisis, which is more than a masculine delight in fantasies of destruction, though it is probably that too, and in face of increasing encroachment on public space by corporate and consumer interests, an agenda for urban renewal is pressing, and entails a re-visioning of how and by whom the form of a city is determined, whether that form is designed to be permanent or to be mutable. This includes interrogating even seemingly progressive or 'liberal' ideas such as the 'liveable city', which is a construction by professionals for their supposition of the benefit of users, its underlying agenda perhaps being the same stability that characterises the rational city, and its content containing a possible contradiction between a desire for a more democratic society and a resort to nostalgic notions of urban history, as if a model of conviviality once existed; but it also arises from ecological concerns and a reaction against the dehumanising effects of consumerism.

A sustainable urban future requires strategies in which the 'edges' of the city – Richard Sennett's difference and diversity – are celebrated, the publicness of urban space reclaimed, dwellers empowered to construct concepts of the city, and in which change is seen as a continuing condition. From the discussion of urban sustainability, and the models of alternative modes of dwelling provided by ecologically and spiritually inspired groups and communities, an agenda is being written to which artists, craftspeople, architects, urban designers and planners might respond. The project is not the construction of Utopia, for most Utopias are authoritarian in the brittleness of control they require to maintain their stasis, but the reconstruction of everyday life and rediscovery of joy in city living.

1

THE CITY

•

INTRODUCTION

This chapter considers changing ideas of the city which are produced by social structures of value and power. It is followed by chapters on the representation of space and its gendering, the role of monuments in constructing national identities and controlling urban publics and the specialist practice of public art, which may extend or challenge the conventions of the monument, be complicit in or resistant to dominant kinds of urban development, open or foreclose arguments about the city.

The chapter summarises arguments from various disciplines which establish that cities are socially produced rather than the product of biological inevitability and that city form is not neutral but imbued with ideologies; it is not an historical survey in the way of Lewis Mumford's *The City in History* and compares only particular cases from the literature of urbanism, such as the orthogonal street plan of the Greek city in the context of Greek material culture, and the construction of images of the medieval city as compensatory for the perceived lack of conviviality in industrial cities; it argues that the idealism of the planned cities of the Enlightenment is an expression of a desire for purification, and that the development of urban planning as a specialist discipline extends this dynamic, not least through the abstraction of the city plan. This distancing view of the city is likened by Michel de Certeau, in *The Practice of Everyday Life*, to the 'gaze'; the chapter also suggests a similarity with Michel Foucault's formulation in *The Birth of the Clinic* of the 'medical gaze'. Both the desire for purity and the analytic viewpoint lead to the fragmentation of the twentieth-century western city; and just as the city is separated into zones, so the process of planning itself is a separation of theory from practice, which leads to a problematic model of the material city underpinned by the conceptual city, and a privileging of that concept. A brief outline of the problem of separating concept and form precedes the main discussion.

VIEWS OF A CITY

An image of a city, such as a postcard view or family snapshot, encapsulates
a relation to the city; this is in part determined by the personal associations
the image may conjure, and in part by the viewpoint from which the city is seen.
A snapshot might be taken in a busy street, so that elements of the city unre-
lated to the subject encroach within the frame, and although characteristic
detailing or materials of a building, or a glimpse of a familiar landmark, might
suggest a particular place, the image gives little idea of the city as a whole, its
configuration, its boundaries; some postcards, in contrast, present a view from a
vantage point outside the city or from a high point within it, lending coherence

FIGURE 9
Trafalgar Square: in
large public spaces
individuals become
a crowd

FIGURE 10 Villiers Street, London: even from the second floor, people become anonymous, and the photographer's is not the only eye watching – note the surveillance camera

to the cluster of buildings, transforming the city into a distinctive skyline. The snapshot suggests an incidental moment within the city, one instant in its diversity; the distanced view re-presents the city as a single idea, as if it might always be that way.

The street offers casual encounters, the possibilities of engagement, the adoption or relinquishing of a personality. From the top of an office tower, or from a viewpoint on a surrounding hill, it is possible to see the city as a map, its pattern and scale of streets, squares and green spaces set out two-dimensionally. From the distant viewpoint, it is not possible to see people as such – from a building of medium height they merge like ants into pools of activity and disperse

FIGURE 11
Manhattan seen
from Battery Park
City – a utopia in
which difference is
invisible

into emptiness or shadow. Even from two levels above the street they become anonymous. De Certeau, looking down on New York from the 110th floor of the World Trade Centre, writes of being 'lifted out of the city's grasp' by a perspective of isolation, a kind of urban gaze which makes the observer 'a voyeur' (de Certeau, [1984] 1988: 92); the viewpoint, then, is spatial, but it implies a difference of value and a different position in relation to power and possession from that of the street (in which everything is around the observer, who becomes part of its interactivity). Perhaps this is one reason why chief executives' offices tend to be on the top floor of corporate buildings – they might like the view, but why?

Concepts and forms

The viewpoint of distance emphasises a unified concept of the city, a pattern which perhaps corresponds towards an ideal city, over the multi-layered collage of constantly changing detail experienced by walking through its streets. The 'invention' of the distant view dates to the fifteenth century and the development of a method of plotting the city from vantage points along its walls, allied to Alberti's system of linear, single-viewpoint perspective;[1] this abstraction, a view as if from the sky, homogenises space in a system of measurement, so that spaces of equal size are of equal value, set out on the blank ground of a sheet of paper.[2] This method, which gives the power of representation to the map-maker whose efforts are aimed at describing a city which exists, suggests a further possibility of fantasising a city, of assuming the power to draw a city which does not exist, that is, of planning a city which can then be made to exist, and is not contingent on the (perhaps disorderly) lives of its inhabitants.

Ideal cities and societies are found in literature, from the book of the *Apocalypse* to Campanella's *City of the Sun*, More's *Utopia*, or Bacon's *New Atlantis* and in Renaissance art, but to realise the idea in actuality is to cross from the spaces of art to those of power, as in Baron Haussmann's re-planning of Paris for Napoleon III and Ceaucescu's new Bucharest (Figure 12), or those of capitalist enterprise, as in the building of New York.[3] The re-configuration of a city introduces, in treating its existing fabric as a contourless ground on which to inscribe a new design, the possibility of a radical break with history. For Napoleon III it was a question of preventing future insurrections, for Ceaucescu of snatching a place in history, but perhaps the urge for a new city derives also from a desire to purge the unclean, abolishing the mess and complexities of the past.

Planning became the dominant approach to city development during the period from the Baroque to the Enlightenment,[4] superseding the *ad hoc* growth of the medieval city and replacing the sacred traditions of archaic cities with the imperatives of secular power and its need for spectacle (Wilson, 1991: 18). Whilst medieval streets were the gaps between buildings, in the Baroque city they became avenues of procession. The form of a city came to represent a concept of order, as it were, illustrated by the built city; because this is a process of plan and execution, it is possible to write of a conceptual city of which the spaces, streets and buildings of the material city are the application. This split of concept and material form is convenient for argument, leading to an interrogation of the concepts underlying city form; but it has the difficulty that the differentiation of concept and material form (and the notion that one underlies the other) is a device which itself reflects the viewpoint of power from which the concept is generated. The alternative to urban idealism, then, is a re-unification of concept and form, which recognises that both users and planners, in a democratic society, have roles in this non-hierarchic process.

FIGURE 12 The utopia of new Bucharest – a public fountain without water

IMAGES OF THE CITY AND SOCIAL VALUE

The relation of cities to social values is proposed from various disciplines, sometimes using cultural metaphors: sociologist Rob Shields writes that 'the city itself can be treated as a representation of the society which constructed and used it' (Shields, 1996: 231); J. B. Jackson sees the re-definition of agricultural space in north America during the nineteenth century as expressive of a pioneer spirit which was 'expansive, free of the past, more and more involved with the transformation of the natural environment' (Jackson, 1972: 30); and historian Robert Fishman suggests suburbia represents 'an archetypal middle-class invention . . . founded on the primacy of the family' (Fishman, 1987: 3). Marshall Berman sees the modern city developer as analogous to Goethe's Faust 'who puts the world on its new path, is an archetypal modern hero' (Berman, 1983: 66),[5] and relates the development of New York masterminded by planner Robert Moses to the optimism of Roosevelt's New Deal. Urbanist Mike Davis describes Los Angeles as a realisation of Brecht's city of Mahagonny, the 'city of seduction and defeat' and 'the archetypal site of massive and unprotesting subordination of industrialised intelligentsias to the program of capital' (Davis, 1992: 19).

Urban catastrophe?

Images of the city in the media and popular culture, even in some academic or professional texts such as *Angst: cartography* (1989) by Mojideh Baratloo and Clifton Balch, or Peter Lang's *Mortal City* (1995), seem increasingly negative, though many are produced by men and may reflect a masculine fascination with destruction. Lebbeus Woods, for example, uses the trope of architecture as war, though his assertion that architecture and development are aggressive rather than benevolent acts may hold (Woods, 1995: 50). Modernism's celebration of the city seems to have given way, via alienation, to despair.

The anonymity of the city of modernity, a literary idea of a new kind of freedom (for men, like Charles Baudelaire, from family and class conventions of behaviour, though not inclusive of economic freedom), is represented by the *flaneur* who strolls through the arcades of Baudelaire's (and Walter Benjamin's) Paris enjoying a distanced identification with the rag-picker[6] and the deviant, or by the introspective isolation of the poet in Rilke's *The Notebooks of Malte Laurids Brigge;*[7] it is also represented in the empirical texts of the Chicago School of Sociology by the hobo (Anderson, 1923), or the Taxi-Hall dancers who danced with men for money (Cressey, 1932), types selected within a cultural framework which they then support, and seen from a bourgeois and masculine viewpoint from which the rag-picker can be romanticised and the equivalent of the *flaneur* is the whore.[8] Fritz Lang's film *Metropolis* (1927), in which the processes of

FIGURE 13 The Winter Gardens at Battery Park City – the creation of new vistas

production enslave an underground work-force beneath the towers of the elite, presents the machine city as a site of alienation,[9] though one in which revolt is still possible; in the 1960s, the educated, urban generation who celebrated free love and protested against the war in Vietnam briefly created enclaves of a new bohemia (mostly now gentrified) or turned to rural escape routes – Woodstock, like Glastonbury, is in the countryside.[10] More recently, as urban culture becomes a diversity of sub-cultures, and the abject who no longer have any relation to the conventions of liberal society are termed an 'underclass', a terminal breakdown of value projected onto Los Angeles in 1999 in Kathryn Bigelow's *Strange Days* (1995) forces the viewer to confront the prospect of a desolate urban future which is credible because there are already so many reports of it in media coverage of 'riots', street crime and serial killing. Architectural critic Mark Wigley argues that the compartmentalisation of violence through media concentration on spectacular collapses of order enables a cultural image of stability to survive (Wigley, 1995: 74), but emphasises the ordinary violence of daily life which goes on despite the absence of cameras, and which suggests a rupture of the idea of the city more powerfully than the isolated but spectacular incident. But though this breakdown of 'law and order' takes a contemporary form, the concept of the undoing of society – a 'last days' – has a long tradition.

St John writes of 'a new heaven and a new earth, for the first heaven and the first earth were passed away, and there was no more sea' (Revelation: 21, 1), on which John Willett comments: 'Ever since Saint John the Divine found himself "in the spirit" . . . the vision of a transformed city, glowing and beautiful, has haunted the human race' (Willett, 1984: 7). St John's image represents a cosmology; the image of 'no more sea' is more than an afterthought, suggesting perhaps an abolition of the 'unconscious'. De Certeau writes, using the term 'scriptural'[11] to mean how society writes its 'script' that revolution 'represents the scriptural project at the level of an entire society seeking to *constitute itself* as a blank page with respect to the past, to write itself by itself' (de Certeau, [1984] 1988: 135). The desire to make new by abolishing the past is taken to an extreme by Le Corbusier:

> Therefore my settled opinion, which is a quite dispassionate one, is that the centres of our great cities must be pulled down and rebuilt, and that the wretched existing belts of suburbs must be abolished and carried further out; on their sites we must constitute . . . a protected and open zone, which when the day comes will give us absolute liberty of action, and in the meantime will furnish us with a cheap investment.
>
> (Le Corbusier, [1929] 1987: 96)

This kind of thinking was applied in Europe, by expert planners and architects, in the clearing of 'slums' and relocation of populations in peripheral housing projects during the post-war years, constituting, perhaps, one of those areas of 'everyday violence' which lead us to question the sustainability of the city.

Future-cities as past cities?

Today, electronic technologies reduce the need for people to share physical spaces, suggesting a new kind of urbanisation available to those who have the technology.[12] Mumford, whose history of the city concerned the enduring elements of walls, houses and streets, foresaw a global city of communications media liberated from consumerism, 'bringing all the tribes and nations of mankind into a common sphere of cooperation and interplay' (Mumford, [1961] 1966: 638) – a new kind of liberal *civitas*.[13] He takes Olympia, Delphi and Cos as examples of integrated religious, political, literary and athletic developments constituting a holistic urban concept from which a co-operative commonality is derived (Mumford, [1961] 1966: 160), defeated only by population expansion and rivalry. Mumford's history speaks of a desire for synthesis and continuity; whilst Jackson, for instance, sees the medieval city as a radical break from the Roman (Jackson, 1980: 56), Mumford writes that 'neither the Roman way of life nor the Roman forms altogether vanished' (Mumford, [1961] 1966: 284). From the contrast between these accounts emerges a vision of a universality also demonstrated in Mumford's image of medieval collectivity; he claims that the craft guilds 'fabricated a whole life, in friendly rivalry with other guilds; and as brothers, they manned the walls adjacent to their quarter' (Mumford, [1961] 1966: 312), but perhaps, sometimes, they also hired thugs to attack their neighbour's customers, beat their wives and killed Jews.

Remote histories offer sites for compensatory models of what is lacking in the present.[14] The image of the 'friendly rivalry' of Mumford's craft guilds is spoilt by Sennett's references to the crime statistics of Paris between 1411 and 1420, when 76 per cent of criminal cases concerned violence against the person, probably connected to alcohol used as a defence against cold and pain (Sennett, 1994: 196); and his vision of the free medieval artisan: 'A city that could boast that the majority of its members were free citizens, working side by side on a parity, without an underlayer of slaves, was, I repeat, a new fact in urban history' (Mumford, [1961] 1966: 313) does not refer to women.[15] Ecologist Herbert Girardet also seeks to construct a compensatory vision of the city; in place of a modernist abolition of the past, he presents a selective past:

> Cities such as Florence, Salzburg, and Prague seem to have been purpose-built for lively interchanges between people. Narrow, human-scale streets contrast with well-appointed public buildings and wide, open gathering spaces
>
> (Girardet, 1992: 118)

which conceals the structures of power and the political and personal violence which took place within these cities. His image of the spatial unity of the medieval town is countered by Jackson: 'The castle, the cathedral, the market, represented separate and distinct centres of activity, and beyond them the town was little more than small groups of flimsy dwellings' (Jackson, 1980: 63); and by Sibley: 'the socio-spatial structure of the city also expressed a wish to erect boundaries to protect civil society from the defiled, even though some writers, notably Lewis Mumford, have idealized medieval cities as socially integrated collectivities' (Sibley, 1995: 52).[16] Girardet, in effect, proposes the absence of the qualities he describes in cities today; whilst a history which constructs the past as an image of present loss is of interest for what it says about the present, nostalgic solutions detract from the ability to address current problems. Wigley claims that the appeal to an idyllic innocence – 'before violence' – is instrumental in 'promoting the violence it appears to be rejecting' (Wigley, 1995: 75). To sentimentalise the past also locates the city in an aesthetic domain, but images of the past which follow the model of art[17] are decried by Rosalyn Deutsche as uncritically reproducing idealist notions of the city (Deutsche, 1991b: 47).

ORDER AND APPEARANCE – THE AGORA AND THE GRID

For Mumford, enduring values were created in Greek society, through its political ideals. It could be asked to what extent such values were available to the slave class on which Athenian democracy depended, or to women confined to houses, but a commonplace image of the archaic city, like that of the medieval city, is of an informal but harmonious mix of dwellings, markets, state buildings and sacred precincts within the protection of a wall,[18] which acts as the organising principle of the surrounding countryside. This could be interpreted as answering biological requirements, or the expression of a cosmological model which is more or less unchanging but is cultural and hence socially produced, or as the product of mutating political and economic histories. The divergence between accounts which claim the ideological neutrality of biology and those which adopt critical methods can be seen by comparing Joseph Rykwert's writing on archaic cities with that of David Harvey on modern cities. Rykwert writes of the rituals associated with the foundation of archaic cities having 'roots in the biological structure of man . . . the formal movement of natural recurrence: day and night, the phases of the moon, the seasons, the changes of the night sky' (Rykwert, [1976] 1988: 194–5); Harvey, on the other hand, relates changes in urban space to economic causes and the need, under capitalism, for constant expansion so that, when an existing pattern blocks further growth, 'the landscape must then be reshaped around new transport and communications systems and physical infrastructures, new centres and styles of production' (Harvey, 1993: 7). For Rykwert the city represents an order which transcends history; for Harvey it is produced (and produces) socially. One form of a social history is Sennett's relation of city form to attitudes to the body in *Flesh and Stone*; another is the interpretation of the orthogonal street pattern in terms of Greek material culture by Indra Kagis McEwen in *Socrates' Ancestor*.

The agora of Athens

Sennett begins *Flesh and Stone* in the agora of Athens, a space in which the bodies and decisions of the male citizen are open to sight:

> To the Ancient Athenian, displaying oneself affirmed one's dignity as a citizen . . . the organisation of the crowd and the rules of voting sought to expose how individuals or small groups voted to the gaze of all. Nakedness might seem the sign of a people entirely at home in the city; the city was the place in which one could live happily exposed, unlike the barbarians.
>
> (Sennett, 1994: 33)

The possibility of nakedness stands for an open society. Accounts of the agora of Athens refer to it as the place where city life is visible and audible,[19] the place of transactions of both goods and ideas, and where artefacts are made,[20] watched over by Hermes, god of trade, the transmigration of souls, and trickery. The inter-relation of trade and politics is supported by myth[21] and archaeology. But, the 'citizens' who mingled in the agora were free men, not women, slaves or strangers, and the agora a gendered space, according to a theory of hot (male) and cool (female) bodies (Sennett, 1994: 40). Here, perhaps, is the beginning of the zoning of the city which, in modern times, dominates its planning.

Sennett's model does not preclude a relation of city form also to the material culture, or ritual dance, a field of tacit rather than intellectual knowledge in which a society's structures of value and thought are implicit. McEwen, re-interpreting the Anaximander fragment,[22] employs a metaphor for the city based in the practice of weaving:

> the notion of a polis allowed to appear as a surface woven by the activity of its inhabitants; the sequential building of sanctuaries over a period of time . . . and the subsequent ritual processions from centre to urban limit to territorial limit and back again, in what can be seen as a kind of Ariadne's dance.
>
> (McEwen, 1993: 81)

And like a well-worn cloth, the city has constantly to be re-made;[23] this metaphor is more than a play of words and equates the grid with the idea of 'appearance' as the heightened visibility of something well-crafted. The orthogonal street pattern which characterises later Greek city form and that of its colonial cities, perhaps introduced in Piraeus to order this zone of strangers, is, then, a likeness of the weft and warp of woven cloth, enabling the city to 'appear';[24] thus the city becomes an artefact, appearing in the (un-crafted) landscape, but not dominating it as an imposition of 'order' on 'chaos'. McEwen writes:

> The Western tradition has been to interpret the skilful embrace of aporia [the immeasurable] revealed through the construction of choros [the measured spaces of the dance floor] and Labyrinth as a question of imposing order on chaos. This is a misrepresentation whose roots may well lie with the Romans . . . those expert pavers of straight roads which for centuries sustained the march of the bearers of the *pax Romana*.
>
> (McEwen, 1993: 60–1)

Hence, the grid of the Greek city is not a mechanistic device for the abstract representation of space or the visibility of power, but a mirror of a cosmos of inter-relation, like the weft and warp of a cloth which could not exist if one were absent, and would not be fine if either were inferior.

PLANNING THE CITY

Separation of one thing from another in order to privilege one – self over environment, city over landscape, theory over practice – characterises the planning of a modern city. The zoning of the modern city, at one time related to fear of corruption by the 'miasma' of the poor (Wilson, 1991: 37–9), strangely links with an increasing concern for free circulation of traffic as a means to healthy growth.

In the ghetto of Venice in the sixteenth century, Turks, Armenians, Germans and Jews were controlled by a night curfew within locked and guarded buildings; this reflected a fear of disease, especially syphilis, spread by 'strange bodies' (Sennett, 1994: 222–51). Urban spatial and cultural boundaries became clear by the eighteenth century, and 'boundary maintenance became a concern of the rich, who were anxious to protect themselves from disease and moral pollution' (Sibley, 1995: 53).[25] The poor were seen with similar distaste to the sick, morally and medically corrupting, leading to an increased separation of working-class and bourgeois districts and location of the former at a distance from civic centres; Sibley relates the zoning of the city to sanitation, and sets out a two-stage model of difference in which the bourgeoisie, having developed a sense of self which excluded bodily residues, could differentiate themselves from the 'smelly' working class; the notions of dirt and disease applied to the poor could then be used in colonial societies to construct images of immigrants (Sibley, 1995).

The city of circulation routes, like veins and arteries, manifest in, for example, a plan of Karlsruhe in the eighteenth century, or the plan for Washington conceived by the French engineer L'Enfant in 1791 (Sennett, 1994: 264–6), echoes Harvey's model of the circulation of the blood published in *de motu cordis* in 1628; Sennett links it to the birth of modern capitalism in Adam Smith's *Wealth of Nations* (1776), taking the free market of labour and goods as a metaphor for blood giving life by circulating freely within the body. The contradiction implicit in this city is that free movement applies only to the male, bourgeois citizen, for whom the coherence of the rationally planned city acts as a defence against its contradictions. Hence free circulation takes place in one realm of the city, but confinement is the condition of whatever might upset the social order. What might block the free flow of the city, like a crowd, is dispersed, and what might puncture its illusion of the ideal, such as vagrancy, is excluded. Better circulation also, by the nineteenth century, enabled more discrete zoning; suburban and metropolitan railways carried the citizen from a suburban home to work in the centre, and servants from poor districts to those of wealth (Sennett, 1994: 332–8). Meanwhile, in Haussmann's Paris, armed force ensured order, using the free-fire zones of the boulevards which also enabled the bourgeoisie to move in and out of the city; the poor were relocated in suburbs and an image of equality was conveyed by the new pleasure grounds.[26] Circulation and compartmentalisation seem, then, to be related.

Exclusion and confinement

The city of circulation is inseparable also from the city of discipline, and the model of free circulation leads in modern development (implemented through bureaucracy rather than autocracy) to the destruction of habitats as 'obstructions to the flow of traffic' in cities which are 'junkyards of substandard housing and decaying neighbourhoods' (Berman, 1983: 307) rather than places where people live. The ghetto, expressing fear of difference, becomes a dominant model of spatial zoning from the eighteenth century, and a particular form of segregation, central to the process of exclusion of the non-productive from the visibility of the street, is the asylum. When the first General Hospital was founded in Paris in 1656, it served as a place into which the vagrant or insane (who were not differentiated) could be removed from public view[27] to a territory of confinement designated as their 'natural' home. The decree founding the General Hospital is more than an administrative reform, and Foucault makes clear that it is not a medical establishment but 'a sort of semi-judicial structure, an administrative entity which, along with the already constituted powers, and outside of the courts, decides, judges and executes . . . jurisdiction without appeal' (Foucault, 1967: 40). The creation of the hospital as a separate 'country' formalises something later ingrained in the modern understanding of the city: in contrast to the 'sinister' spaces of the 'heath' – given to witches and outlaws – the city of the Enlightenment is clear: light, ordered and predictable, a set of masculine traits. Street patterns of the eighteenth and nineteenth centuries 'clarify' the accumulations of the medieval city by cutting through them, making open spaces for shows of state power. Elements of the population seen as impurities are by the nineteenth century confined in a range of institutions with common architectural languages: the asylum, the workhouse, the infirmary, the prison and the school, whilst the productive poor are incarcerated in factories.

Not only the deviant, vagrant and insane but also the dead, who deflate the notion of perfection in their putrefaction, and the mess of the living, are, from the eighteenth century, excluded from the visible and sensible city. Ivan Illich writes:

> People then not only relieved themselves as a matter of course against the wall of any dwelling or church; the stench of shallow graves was evidence that the dead were present . . . Universal olfactory nonchalance came to an end when a small number of citizens lost their tolerance for the smell of corpses. . . . In 1737 the French parliament appointed a commission to study the danger that burial inside churches presented to public health.
>
> (Illich, 1986: 50)

The invisibility of waste had already been proposed by Alberti in the sixteenth century;[28] Wigley notes: 'social order needs to be cleansed of the body. Architecture is established as such a purification . . . The body itself emerges as a threat to the purity of space' (Wigley, 1992: 344). Illich, too, states that the exclusion of odours and the waste products of the body from the city is not simply a response to medical problems but that philosophical and juridical arguments were used amidst a general fear of 'miasma' (Illich, 1986: 50).

The dynamic of exclusion and confinement has taken new forms in modernity. In *Conscience of the Eye*, Sennett writes of modern north American cities in relation to a puritan characterisation of the environment:

> The cultural problems of the city are conventionally taken to be its impersonality, its alienating scale, its coldness. There is more in these charges than is first apparent. Impersonality, coldness and emptiness are essential words in the Protestant language of environment; they express a desire to see the outside as null, lacking value.
>
> (Sennett, [1990] 1992: 46)

In *Flesh and Stone*, Sennett refers to the bleakness of modern cities related to a concept of the environment in which the body is excluded (Sennett, 1994: 15). Perhaps it is also helpful to take the psychoanalytic model of defensive splitting[29] as a way to understand this bleakness – just as the infant splits off the 'bad' mother, so the urban dynamic of exclusion and confinement is a splitting off of the diversity of urban life which threatens its purity, but leaves it a wasteland. In the literature of psychoanalysis, splitting is described as an unconscious process: 'the ego can stop the bad part . . . contaminating the good part, by dividing it, or it can split off and disown a part of itself. In fact, each kind of splitting always entails the other' (Mitchell, 1986: 20). That which is different, which fails to conform to the consensus of rationality, or the implicit logic of inwardness, is rejected to construct a false security requiring drastic measures to enforce it. City zoning is a form of purification, of setting up a rigid model which excludes differ-ence, of splitting off the unacceptable other, the dirt.[30] The patterns of thought which underlie this history are reflected in attitudes to the flowing of water, the burial of the dead and disposal of waste, and the treatment of vagrancy, but no less in the zoning of the modern city. They are also echoed in the signs amongst the carefully regulated shrubs of Battery Park City which discourage the feeding of pigeons and squirrels on the grounds that it might encourage rats (Figure 14).

URBAN PLANNING

The city of exclusions where elements like the poor are repressed into marginal spaces (like dark thoughts into the unconscious), was the dominant idea of the city when, in the early twentieth century, town planning became a specialist discipline. Single-use zoning, characteristic of urban planning in the twentieth century, separates the spaces of domestic life (the dormitory suburb) from those of government (the civic centre), labour (the industrial estate or business park), leisure (the multiplex, the sports centre, the museum or heritage district) and consumption (the mall). The house, too, is divided into rooms of a single function, and the principle that physical space should be allocated predetermined use remained an unchallenged orthodoxy of urban planning until Jane Jacobs' *The Death and Life of Great American Cities* (1961). But another idea of the city, projected back onto past cities, is that they are places of unplanned mixing.

Georg Simmel, in *The Metropolis and Mental Life* and other writing, sees urban sociation – the way people in a city relate to each other – as having just such an informality, whilst the constant stimulation of the city means that the citizen becomes inured to it; Simmel, whose sense of modernity was influential on the emerging Chicago School of sociology, sets the modern city apart from both rural settlements and from earlier cities less affected by the money economy, distinct spatially and temporally, 'the urban' itself becoming a defining characteristic of modern life.[31] Studies in urban ethnography identified city types, such as the hobo, rather as Baudelaire identified the *flaneur*, and Walter Benjamin, in his unfinished Arcades Project listed several types including the salesman, the collector and the prostitute,[32] as representations of modern urbanism. In the 1920s, the Department of Sociology at the University of Chicago, following an agenda for

FIGURE 14 A notice at Battery Park City – wildlife becomes abjected. So should the pigeons and squirrels just die?

FEEDING PIGEONS AND SQUIRRELS LEAVES FOOD FOR RATS

research derived from Robert Park's essay *The City*, produced studies of land use and social differentiation. E. W. Burgess, for whom 'the outstanding fact of modern society is the growth of great cities' (Burgess, [1925] 1972: 117) conceived, in his 1925 essay *The Growth of a City*, a mapping of the city in concentric zones, using a model from natural science[33] which ignored features of the underlying landscape; this theoretical model could be applied to any city, and still dominates the iconography of the subject (Sibley, 1995: 152–3), influencing geographers as well as sociologists.[34] It included a central business district identified with the Loop in Chicago, an area of transition into which new industries moved, a zone of working-class housing, outer residential suburbs for middle-income earners, and beyond that a commuter zone for the wealthy.[35] It seemed to Burgess as if the city was subject to waves of 'invasion', again borrowing a model from plant ecology (Burgess, [1925] 1972: 120); but the language of the natural sciences is used to situate urban development beyond human agency. Burgess and the Chicago School take their diagrams as rationalisations of research data, descriptions from which planning decisions could be made, and a means to make a specific form of city development seem inevitable, prescriptions which tend to ensure future data agree with the initial descriptions. The ring model is, however, ideological, mirroring a model of an ever-expanding economy and centralised power.[36] Burgess goes on to ask what is a 'normal' rate of expansion in relation to biological processes of metabolism (Burgess, [1925] 1972: 122), from which he normalises a measure of disorganisation through urban immigration, leading to the production of zones of transience, such as 'hobohemia', and a zone of 'bad lands' with vice and poverty 'always' found around central business districts.

The work of the Chicago School was a structured enquiry into moral codes and conventions which explained apparently haphazard forms of human action (Savage and Warde, 1993: 12). In modern urban society, they maintained, the ties of sociation are more complex and concealed, more mutable, than in rural societies, and ties based in shifting occupations and neighbourhoods replace those of family; the Chicago sociologists categorised these new relations, whilst accepting uncritically the expansion of capitalism. Although they were seen as reformists, their contribution establishes the sociologist in a remote location, like de Certeau looking down from a high building but without the critique which enables de Certeau to know what he is doing. Duncan concludes: 'Burgess' dream of the happy unity of the natural and human sciences pushes him to transform Chicago into "the city" in ways which stretch credulity' (Duncan, 1996: 265).

A relation between sociology and the discipline of urban planning was established by Louis Wirth in his essay *Urbanism as a Way of Life* (Wirth, 1938). Wirth's point of departure was the opportunistic marginality he identified as a difference between urban and rural sociation, in context of the size, density of population and heterogeneity of the modern city;[37] in the metropolis of lonely individuals, human relations become segmentalised, transitory and largely anonymous. The specialisation of work produces a complex model of segregation which makes it difficult to generate a sense of community when the links people make are not based on place or family and most of the people the urban dweller encounters are strangers: 'Today all of us are men on the move and on the make, and all of us by transcending the bounds of our narrower society become to some extent marginal men' (Wirth, 1938, cited in Smith, 1980: 9). Wirth's solution is rational consensus and scientific objectivity in which social cohesion is engineered by urban planners operating in zones corresponding, for decisions on matters such as housing, transport and taxation, to the whole metropolitan region – regional planning.

Wirth concentrated on the function of size in cities and the differentiation of the urban from the rural, and ignored the impact of the money economy (of prime importance to Simmel) as the site of alienation within urban life;[38] Savage and Warde state several objections to Wirth's work: 'inquiries found communities in the city and conflict in the countryside [and] the diversity of group cultures challenged the idea that there was one dominant way of life' (Savage and Warde, 1993: 99). But a basic problem with Wirth's argument is that there is no clear model of how a consensus might operate in a way which empowers urban dwellers as co-designers of their habitat; more often it is the bureaucracy of city authorities which acts to represent this consensus without any means of constructing it. John Forester proposes an application of the critical theories of Jurgen Habermas: he summarises Habermas's communications theory of society as stating that structures designed to communicate also convey ideology and obstruct freedom through displacement, so that political questions are presented as technical matters for choices on the level of detail, and many implications of development which might tinge the developer's aura with self-interest are unstated in public debate; a process of social criticism and mutual understanding addresses this disinformation to form a more open consensus (Forester, 1987a: 204). This raises several problems, in particular: the need for freedom of information and the acquisition by professionals of a vocabulary through which to discuss urban planning and design with urban publics.

THE 'PLANNING GAZE'

Grass-roots planning is still in its infancy; most decisions depend on the conventional devices of the city plan. The viewpoint of height is crucial to conventional urban planning and represents power; Haussmann's urban geometers looked upon the city from tall towers of wood, enabling them to see it like a plan on which to inscribe the boulevards, and Park wrote: 'make of the city a social laboratory or clinic in which human nature and social process may be conveniently and profitably studied' (Park cited in Savage and Warde, 1993: 18). If the city is like a clinic, then the sociologist works with the same 'gaze' as the physician. Foucault, in *The Birth of the Clinic*, writes of a 'constant gaze upon the patient, this age-old yet ever renewed attention that enabled medicine not to disappear entirely with each new speculation, but to preserve itself' (Foucault, 1973: 54), and of the moral problems which arose when the poor who sought aid at the hospital were used as vehicles for medical training in the clinic. The patient 'was now required to be the object of the medical gaze . . . what was being deciphered in him was seen as contributing to a better knowledge of others' (Foucault, 1973: 83), just as those evicted from apartment houses demolished for new expressways contributed to a better traffic flow; if it is the purpose of the doctor 'to isolate features, to recognise those that are identical and those that are different, to regroup them, to classify them by species or families' (Foucault, 1973: 89), then perhaps the planner does something similar and perhaps zoning is an equivalent act of classification, making the coherence of marginalised sociation invisible to those who propose to clear areas they regard as derelict.[39] The analogy is speculative, but the model of the medical gaze seems similar to that of the 'planning gaze'; crucial to the argument is the urban plan as a 'representation of space'.[40]

The concept of a city expressed in its plan and structures of administration is a pervasive influence on social behaviour, affecting how people of different genders, ages, classes, and ethnic origins mix or remain segregated. Yet it is the detail of material form which is more often offered for 'consultation' by urban planners and designers, and discussed in the definition of what Kevin Lynch calls 'good city form', rather than the question of what kind of city meets the needs of which publics. Lynch writes, at the beginning of a chapter on 'City Models and City Design', that 'Design decisions are largely based on models in the head of the designer' (Lynch, [1981] 1984: 277) which suggests a conceptual level of operation, yet limits its scope to the mind of an individual designer, as if designers functioned in a realm of free choices, without questioning received conventions, structures of power or the relative roles of professionals and users in determining city form. He makes several observations on the axial networks of the Baroque city: 'a simple, coherent idea that can be rapidly employed in a great range of complex landscapes', 'a strategy for the economical application of central power', 'a workable strategy for opening up an existing circulatory maze', and notes its defect in causing traffic congestion at the nodal points (Lynch, [1981] 1984: 281–3), all of which are statements about the mechanisms of the city, as if such mechanisms (and statements) could be ideologically neutral. Geographer John Rennie Short, however, emphasises that city form is not static: 'The city is a metaphor for social change, an icon of the present at the edge of the transformation of the past to the future' (Short, 1991: 41), from which the question arises 'in whose interests, does it change?'; Wigley states: 'we know that the physical form of the city is radically changing all the time . . . We also know that people's relationships to the city change and, therefore, they change. That brings up the issue who is the "we" that defines the city' (Wigley, 1995: 71) – a question posed in another way by sociologist Sharon Zukin: 'To ask "Whose city?" suggests more than a politics of occupation: it also asks who has the right to inhabit the dominant image of the city' (Zukin, 1996: 43). Elizabeth Wilson suggests an answer: 'We will never solve the problems of living in cities until we welcome and maximise the freedom and autonomy they offer and make these available to all classes and groups' (Wilson, 1991: 9). If urban dwellers are to be empowered to challenge the values of urban development, then the relation of urban form to social value, the unpacking of the underlying concept of the city, is the point of departure.

2

SPACE, REPRESENTATION AND GENDER

•

INTRODUCTION

The previous chapter concerned the fragmentation of the modern city and the separation of concept and form. This chapter considers Henri Lefebvre's comple-mentary formulation of the 'representations of space' and 'representational spaces' in *The Production of Space*, and the relation of 'representations of space' to the interiority of Reason; and issues of gender in relation to public space and public art, with reference to the work of geographer Doreen Massey, and through a provocative image given by Luce Irigaray as a point of departure for a resistance to the exclusion of women from the public realm. But it begins with a return to the problem of representation – noted in the previous chapter's discussion of viewpoints from which a city can be experienced or observed – through Barthes' *Empire of Signs*, which refuses a transparent relation between a text and its referents.

It is helpful to consider the question of representation because an assumption is often made that the signifiers used in everyday representations, such as the symbols on a map, have a transparent relation to the things they signify, though a reading of codes of representation often depends on other kinds of information, or on memories, and in everyday urban situations people do not consider the 'critical and even pessimistic study of the non-resemblance' (Shields, 1996: 229) between representation and life; they simply doubt the neutrality of 'information' given in a newspaper headline, an advertisement or a government report.

For Barthes, the text is a set of signifiers without signifieds; this confines him within his text, his political artifice, if he has one, being irony. The contention of this chapter, however, supported by several writers on urban issues, is that the value of theory is at least in part in its application and of criticism in its construction of possible futures of greater freedom.

FIGURE 15 A corporate atrium in Manhattan is called 'public space', but remains corporate space

THE PROBLEM OF REPRESENTATION

The forms of representation of urban space, including the seemingly technical or functional such as the planning specification or public transport map, and the incidental, such as the postcard view or street sign, convey ideology as well as information.[1] A drawing such as *Perspective of an Ideal City* attributed to Francesco di Giorgio,[2] is situated in an aesthetic territory, a universal image to which no city corresponds but to which any city might be made to conform intellectually; but a plan showing a city's roads whilst giving no information on its ethnic mix or its zones of wealth and poverty is a no less selective, if commonplace, representation.[3] Sociologist Rob Shields writes:

> New York is metaphorically represented as 'the Big Apple', but other representations such as maps are understood to be realistic images for every city. In everyday life we fashion and receive countless representations. Of course we all realize that a totally accurate representation . . . is impossible
>
> (Shields, 1996: 228)

and J. B. Jackson that: 'The image of the contemporary city, the sign or logo which all of us know how to interpret, is a blend of cartographic abstraction and aerial view' (Jackson, 1980: 55). Art historian and critic of public art Patricia Phillips argues that 'The map is an iconic representation that leads to illumination, but it is enlightenment that does not necessarily simplify or explain . . . the cartographer both replicates and conceives' (Phillips, 1989: 5). Map-making, or land-registry, despite their functionality, are, like the postcard view, ideological statements.[4] And Lefebvre writes:

> To study the problems of circulation, of the conveying of orders and information in the great modern city, leads to real knowledge and to technical applications. To claim that the city is defined as a network of circulation and communication, as a centre of information and decision-making, is an absolute ideology; this ideology proceeding from a particularly arbitrary and dangerous reduction-extrapolation and using terrorist means, sees itself as total truth and dogma.
>
> (Lefebvre, 1996: 98)

Images of a city, then, whether literary, graphic or cartographic, are authored interpretations projecting a conceptual model (devised in a professional space) onto, or claiming its derivation from, the built city, to articulate or condition thought about cities.[5]

The conventional city plan adopts a viewpoint as if in the sky above the city, looking down from god's eye, the position of power from which alone such an all-knowing representation can be conceived – de Certeau's voyeurism,[6] a matter of repression and desire – 'this lust to be a viewpoint and nothing more' (de Certeau, [1984] 1988: 92), and perhaps it is a form of representation which makes the city (as if) unreal to those who see it this way professionally, whether modern planners of freeways or Haussmann's urban geometers looking down from their towers on a closely knit (but to them worthless or subversive) pattern of alleys; it is also a language which leads to the perplexity of those unfamiliar with it when confronted with architectural plans and asked for their 'views'.

An empire of signs?

Roland Barthes' *Empire of Signs*, a work produced following the 1968 uprising in Paris, refuses to accept transparency in the relation of representation to actuality, utilising a number of verbal and visual images comprising a 'system . . . which I shall call Japan' (Barthes, 1982: 3). The role of codes of recognition in interpreting imagery is well rehearsed; Wigley states that 'Buildings are mechanisms of representation – systems of organising things – and in that sense they are political constructions' (Wigley, 1995: 72) and Norman Bryson has given an account of the semiology of visual interpretation (Bryson, Holly and Moxey, 1991, Chapter 2);[7] but Barthes refuses to attach his signifiers to signifieds. He writes that grid cities such as Los Angeles produce uneasiness through a lack of centre, that 'the West has understood this law only too well; all its cities are concentric' (Barthes, 1982: 30),[8] and describes Tokyo as a city turning around 'a site both forbidden and indifferent, a residence concealed beneath the foliage . . . an opaque ring of walls, streams, roofs, and trees whose own centre is no more than an evaporated notion' (Barthes, 1982: 30–1). This is, or is not, the Japanese city with its absence of street names, a product of Japanese culture as much as the cuisine on which Barthes also writes; but for him 'Orient and Occident cannot be taken here as "realities" to be compared and contrasted historically, philosophically, culturally, politically' and all that can be addressed 'is the possibility of a difference' (Barthes, 1982: 3);[9] the possibility is glimpsed by placing sign by sign rather than 'mining' into a level of signification 'below' the sign.[10] Is, then, Barthes' writing like Italo Calvino's – who sets up a web of differences, a system of categories and numbers, in *Invisible Cities* and writes of Eudoxia that its true form is preserved in a carpet 'laid out in symmetrical motives whose patterns are repeated along straight and circular lines' (Calvino, 1979: 76) which spectators become convinced are to be found in a city which looks unlike the carpet, but also of Olivia that 'the city must never be confused with the words that describe it. And yet between the one and the other there is a connection' (Calvino, [1972] 1979: 51) – despite that Barthes uses the names Japan and Tokyo, which he visited, and Calvino is a novelist?

A self-contained discourse, as claimed by Barthes – 'I can also – though in no way claiming to represent or to analyze reality itself (these being the major gestures of western discourse) – isolate somewhere in the world (*faraway*) a certain number of features' (Barthes, 1982: 3) – may be, outside its academic base, a model of passivity bordering despair;[11] it is described by Berman as a strand of modernism in the 1960s: 'the one that strives to withdraw from modern life' and linked by him to Clement Greenberg's reductionist theory of modern art (Berman, 1983: 29). Berman, who experienced the destructive impact of the construction of the South Bronx Expressway, legitimises his own resistance acts, such as blocking traffic, with the words 'we were trying to open up our society's inner wounds, to show that they were still there . . . that unless they were faced fast they would get worse' (Berman, 1983: 328). Diane Ghirardo calls for 'substantive reform in labor laws, in education, in resource allocation, and in a system of justice that is far from color blind' in response to the 1992 Los Angeles riots (Ghirardo, 1995: 101), while Jon Bird and his co-editors of *Mapping the Futures* write 'It is hard not to feel numb, or powerless, or apocalyptic' (Bird *et al.*, 1993: xv) but also 'The purpose of these essays is to try to gain some purchase on present changes and to extrapolate from them possible futures' (Bird *et al.*, 1993: xv), and Lebbeus Woods calls for responsibility: '[architects] should make their choices consciously . . . accepting personal responsibility for what they do' (Woods, 1995: 53). There seems, then, an urgency to which the response of a closed system is inadequate – James Duncan says of Barthes that his writing 'leaves us somewhat bemused as to his allegedly political intentions, for we know that there are millions of citizens of that city whose lives will remain forever untouched by the writings of one self-absorbed, hyper-intellectual Frenchman' (James Duncan, 1996: 265) – perhaps this was the case in 1968 when Barthes began to think about *Empire of Signs* and Sartre took to the streets, joining the demonstrations of students and workers.

But on what is intervention founded, if not a system of thought which will replicate itself in (intellectual or practical) solutions, or the images with which metropolitan dwellers in a media age are assailed? News coverage, immediate yet safe, crossing the boundaries of public and private domains and neutralized by being mixed with consumer advertising, provides another form of the problem of representation, so that the images of the media constitute another 'system called . . .' which contributes, with the seeming inevitability of city development, to a sense of disempowerment.

Strategies of empowerment are therefore a means to a re-visioning of urban living; but a critique of the representations of space utilised in the conceptualisation of the city is also necessary. This, because the spaces and concepts of the city are gendered, raises (at least) two questions: 'are the ways in which space is conceptualised inherently part of a masculine framework which both privileges abstraction and excludes women from the public realm?' and if so, 'by what strategies can this disequilibrium be addressed?'

THE REPRESENTATIONS OF SPACE

The twentieth-century city is a site of desolation,[12] a wasteland. Does the city's bleakness reflect a cultural attitude founded in masculine traits such as a refusal of feeling, and does the process of planning and design originate in a masculine space of interiority in which a refusal of feeling becomes possible?

Bleakness and interiority

Richard Sennett claims, in *Conscience of the Eye*, that the cold aridity of the modern city reflects a Protestant attitude in which the drabness of the environment encourages a required inwardness.[13] Protestantism affirms patriarchy through its emphasis on the powerful father, a refusal of the feminine seen not least in the marginalisation of the Mother of God, and in the rhetoric of sin and redemption so often translated into hate for self and others. Another image of stern and unforgiving, unfeeling masculinity is seen, coincidentally, in David's painting of Brutus stoically refusing emotion as the lictors carry in the corpses of his sons. David represents two spaces in his picture, divided by compositional devices such as columns and drapes: Brutus, a man of public duty, is ensconced in the space of reflection, like a man in his study (or studio), whilst the women inhabit a domestic interior in which they shriek bodily, flailing the air with their arms. The city, the republic, is safe, the image seems to say, in the hands of unflinching men like Brutus.

In allegorical sculpture, as in verbal language, the city, like nature, is represented as feminine: the city as mother and protectress,[14] its walls a girdle to protect the space of citizenship, as the *Statue of Liberty* is 'mother of exiles', or as prostitute, or Queen.[15] Yet there is a contradiction: the gendering of spaces in the (feminine) city reflects an order which divides a privileged (masculine) public realm, in which is constituted 'the city', from a devalued (feminine) realm of the house, in most modern cities located in a suburb. The city is the object of an abstracting (masculine) gaze, its ground a space free of value (a feminine 'other') and ready, like a bride, for the (masculine) inscription of grand narratives deriving from the abstract thought of interiority, narratives, that is, produced in the secluded spaces of philosophy, or architecture. A product of this relation of blankness and projection is compartmentalisation, a deprivation of diversity, as in the single-use zoning discussed in the previous chapter.

The production of space

There is a further set of contradictions between the way space is conceptualised and the diversity of things which go on in the social spaces of the city. Henri Lefebvre, in *The Production of Space*, argues that, whilst history is bound up with the *time* in which forces clash, *space* is constructed, in modern western thought, as an illusory transparency. Space is also, according to geographer Doreen Massey, categorised as one side of a duality with gendered associations:

> There is a whole set of dualisms whose terms are commonly aligned with time and space. With time are aligned History, Progress, Civilisation, Science, Politics and Reason, portentous things with gravitas and capital letters. With space on the other hand are aligned the other poles of these concepts: stasis, ('simple') reproduction, nostalgia, emotion, aesthetics, the body.
>
> (Massey, 1994: 257)

Lefebvre is concerned to reclaim space as experiential *as well as* conceptual, to turn, perhaps, the contradiction into a paradox, and he asks if there is a logic to space, whether it has limits, and where it meets its first obstacle, such as a 'residue resistant to every analytic effort' (Lefebvre, [1974] 1991: 292–3); from 'peasant' society he takes images of the body, the home and the village with its fields, beyond which is strangeness, and the beginnings of a spatial organisation of the included and excluded which can be both tacitly known and represented, both concrete and social, mental and abstract. The problem is that academic disciplines – he names anthropology – privilege the mental/abstract; hence, for example, Lévi-Strauss can discuss the family and social relationships in a logical way without mentioning the erotic aspect of sexuality, and objects are reduced to signs, so that an inert space which 'corresponds to the Cartesian model (conceiving of things in their extension as the "object" of thought)' becomes 'the stuff of "common sense" and "culture"' and a 'picture of mental space developed by the philosophers and epistemologists thus became a transparent zone, a logical medium' (Lefebvre, [1974] 1991: 297). This logical medium is where the spatial organisation of ideas takes place, as in a hierarchy such as the 'branches of knowledge'. Lefebvre writes: 'The most pernicious of metaphors is the analogy between mental space and a blank sheet of paper upon which the psychological and sociological determinants supposedly "write"' (Lefebvre, [1974] 1991: 297).[16] He relates city zoning to Cartesian rationality and states that it is responsible for the fragmentation of urban society through the agency of a bureaucratically decreed image of coherence (Lefebvre, [1974] 1991: 317).

Massey draws attention to Lefebvre's concern with the gendering of conceptualised space and the spaces of modernity,[17] and cites his critique of Picasso's work as

> a dictatorship of the eye . . . aggressive virility in which Picasso's cruelty toward the body, particularly the female body, which he tortures in a thousand ways and caricatures without mercy, is dictated by the dominant form of space, by the eye, by the phallus – in short, by violence.
>
> (Lefebvre, [1974] 1991: 302, cited in Massey, 1993: 182–3)

Perhaps, then, the space of representation is a split-off aspect of reality, reducing space to a value-free commodity, like paper.

Lefebvre develops a precise terminology to distinguish the space of experience and everyday life with its embodied interactions, and that of abstract thought. This is the key part of his theorisation: The *spatial practice* of a society is its experience of space through what it collectively does: the use of gateways, methods of transporting goods and people, and marking the boundaries of property; *representations of space* are the perceptions or conceptions of space which use signs and codes to enable a common language of space – 'conceptualised space, the space of scientists, planners, urbanists, technocratic subdividers and social engineers' (Lefebvre, [1974] 1991: 38) – and involves, for instance, making maps, verbal and intellectually formulated devices, hierarchies of spaces, and the hierograph of the city, a cross in a circle; *representational spaces* are the lived and felt spaces of everyday life known through its associated images, and involve non-verbal communication, appropriation, rituals, riots, markets and other aspects of life in the street.[18] The formulation is a model, which Lefebvre sees as a whole, of: experience; perception and conceptualisation; imagination and feeling.

If perception takes coded forms which become concepts, then it seems that in western society the conception of space – the *representations of space* – has come to dominate experience. Lefebvre locates the beginning of this in the developing economic relations of Tuscany in the thirteenth century, when the oligarchy of merchants and burghers began to organise the countryside in a new way; serfs were replaced by *metayer* with an interest in production and living in houses (*poderi*) grouped in a ring around a mansion which the proprietor would occupy from time to time. This enabled the bourgeoisie to set up a two-way trade in food and goods; and it established a new experience of space. He describes the alleys of cypress trees between mansion and house, and notes the simultaneous appearance of the urban piazza and subsequent development of a new urban environment and perspectival way of depicting it,[19] whilst representational space is not abolished:

> This is not to say that . . . townspeople and villagers did not continue to experience space in the traditional emotional and religious manner . . . an interplay between good and evil forces at war throughout the world, and especially in and around those places which were of special significance for each individual: his body, his house, his land.
>
> (Lefebvre, [1974] 1991: 79)

Whilst McEwen's relation of the grid to weaving derives from the practices of material culture (McEwen, 1993), the piazza and perspective derive from a new conceptualisation of the city at the end of the Middle Ages, produced by its economic relations.

Gendered representations

In Alberti's writing,[20] we encounter the framework in which the theory of perspective is set. He begins with an open window 'through which I view that which will be painted there' (Alberti, *Della Pittura*, fol. 124 *v*, (ed.) L. Mallé, 1950: 70, cited in White, [1957] 1972: 122). His notion of harmony is a subordination of parts to a whole, elaborated in the rules of composition and given a moral dimension in writing on the family:

> You know the spider and how he constructs his web. All the threads spread out in rays. . . . Let the father of a family do likewise. Let him arrange his affairs and place them so that all look up to him alone as head, so that all are directed by him.
>
> (Alberti, *Della Famiglia* III, 206, cited in Wigley, 1992: 339)[21]

Alberti's model establishes the idealised representation of space as a precondition for a further series of intellectual developments, so that, to state simply, the blank ground for perspective drawing is also the white paint of the walls of the gallery for contemporary art, the blank ground of the city plan or designer's computer screen. Lefebvre gives a cameo of the architect at the later stage of this development:

> the architect ensconces himself in his own space. He has a *representation of this space*, one which is bound to graphic elements — to sheets of paper, plans, elevations, sections, perspective views of façades, modules, and so on. This *conceived* space is thought by those who make use of it to be *true*, despite the fact — or perhaps because of the fact — that it is geometrical
>
> (Lefebvre, [1974] 1991: 371)

linking the representations of space with the interiority of the study, a masculine domain in domestic architecture since the fifteenth century.[22]

REASON AND INTERIORITY

Where does this masculine abstraction which de-values exteriority arise? In the space of a Bavarian stove[23] in the seventeenth century, a method of argument was formulated which excised the body as the unreliable if seductive site of sense impressions:

> Now I will close my eyes, I will shut my ears, I will turn away from all my senses, I will even efface from my thought all images of bodily things. . . . Or at least because of the difficulty of doing this I will deem them to be empty and false.
>
> (cited in Benjamin, 1993: 46–7)

Empty and false is without value: the impressions of the senses, the texture of the lived spaces of the city, the physical and psychological spaces around the body, all waste. Descartes – whose text this is – was aware of the religious debates of his time, spending part of his life in the Netherlands, and of Harvey's publication of the circulation of the blood, and the text might, in its desire to efface sense impressions and by implication the body which is their site, almost describe contemplative practices based on meditation treatises in the emerging Protestant tradition; in these, as in the Jesuit models on which they are, despite the conflicts of religion, based, the soul is refined through exercises. For the Jesuit these use (imagined) impressions from all the senses, as if the meditator is present at the holy event,[24] but in other (seventeenth-century) sources reason intercedes to construct a representative image. The text suggests this more brittle world, in which the glint of light on a glass or on a peel of lemon, or the bloom of a grape in a Dutch seventeenth-century still life painting, is but vanity, a heap of illusions, dry dust or just paint.

The space of Descartes' study is a space for interiority.[25] Such reflection, taken to an extreme, might allow a fancy that the mind is disembodied or the body made of glass.[26] Bodies inhabit space, desiring and suffering in its substance; minds construct inviolable and abstract spaces, as a defence against suffering. Using the method of doubt, Descartes sees the body, lacking epistemological proof of its existence, split off from mind and replaced by representations, turned into text. His philosophical project, presented in the tone of an armchair talk,[27] is the construction of a tabula rasa. Andrew Benjamin, in *The Plural Event*, emphasises its displacement:

> dualism not only demands a radical separation between mind and body . . . [the] supremacy of the mind and the subsequent reintroduction of the body are themselves premised upon this founding separation. The body is at first to be denied and then reintroduced afterwards. . . . The body will have become an object of thought.
>
> (Benjamin, 1993: 46)

It is not only the modern disciplines of science which extend from this position, but also the planner's gaze which purifies the city; the rise of professions, such as medicine, architecture and art, based on an exclusive, specialised knowledge, and which tend to be male-dominated, is one of the prominent characteristics of the modern world.

EXCLUSION

The exclusion of the feminine, and exclusion of women from the public realm, mirrors the purification of the Enlightenment city and operates on two levels:

FIGURE 16 Jim Dine's bronzes referencing Venus seem like a foil to the masculinity of modernist architecture, their curtailment perhaps a kind of disempowerment, a subjection of the feminine to the male gaze

women are marginalised in the professions and public spaces of modern urban life – denied the public realm; and representational spaces are marginalised by the dominant conceptualisation of the city.

Illich describes the exclusion of odours, such as the smell of the dead, from the city (Illich, [1984] 1986: 50–7); Sibley refers to the link between purification and abjection: 'we can anticipate that a feeling of abjection will be particularly strong . . . difference will register as deviance, a source of threat to be kept out through the erection of strong boundaries, or expelled' (Sibley, 1995: 78), and philosopher Gillian Rose, in discussion of the absence of bodies from the methods of time-geography, relates purification to masculinity:

> The history of the white masculine bourgeois body in Euro-America can be told in terms of a series of exclusions . . . Bakhtin and Elias have traced that body's loss of vulgar and feminine orifices and excretions from the seventeenth century onwards; the civilized body was one with limited and carefully controlled passages between its inside and outside.
>
> (Rose, 1993: 77)

Which made the civilised body unlike the female body with its processes of menstruation and childbirth. There is a long history of the differentiation of male and female bodies and their assignment to the public and domestic realms. In *Flesh and Stone*, Sennett describes the Greek notion that the sex of an infant is determined by the level of heat in the womb, male offspring being those which are well-heated, in life inhabiting hot bodies, whilst those of women are cool. On the principle of like to like, men occupy the sunlit spaces of the agora and the pnyx, the spaces of visibility, transaction and decision, whilst women (and slaves), in their cool, clothed bodies, occupy the enclosed spaces of domesticity.[28]

Women and urban space

The exclusion remains despite a change in the model of physiology; Massey, in *Space, Place and Gender*, writes of her childhood visits to Manchester, passing numerous football and rugby pitches – wide spaces entirely given over to boys – and a later visit to an art gallery in which many of the pictures were of naked women painted by men, women seen through the masculine gaze (Massey, 1993: 185–6). Wilson begins *The Sphinx in the City* with a memory of childhood visits to London: 'Every excursion we made together was an immense labour, a strenuous and fraught journey to a treacherous destination: we waited for buses that never came, were marshalled into queues that never grew shorter' (Wilson, 1991: 1). Rose, too, reminds us that urban space is gendered: 'Women know that spaces are not necessarily without constraint; sexual attacks warn them that their bodies are not meant to be in public spaces' (Rose, 1993: 76). Urbanist Louise Mozingo, in an investigation of the uses of public space by men and women, lists differences in the received psychology of urban space, such as women having smaller personal spaces than men, finding crowded situations less stressful than men, and in groups having smaller territories than groups of men (Mozingo, 1989: 40). She cites research (by women) suggesting that women are more likely to move out of other people's way than men, that they are more often touched, and that whilst men, if accosted, tend to be asked for information, women are sexually intruded-on. Mozingo concludes her general argument by saying, 'The small numbers of women in downtown open spaces suggests these environments do not provide women with a range settings that make them psychologically comfortable' (Mozingo, 1989: 41), but does not interrogate the cultural frameworks within which the behaviour of which she writes takes place. Her observations of two open spaces in San Francisco include that men were found to prefer the unpredictability of spaces opening onto the street, whilst women looked for relief from environmental stress in areas which encouraged sociation: 'women may be seeking "back yard" experiences and men be seeking "front yard" experiences' (Mozingo, 1989: 46), which leads her to criticise aspects of W. H. Whyte's *Social Life of Small Urban Spaces* as, in part, a masculine model (from urban sociology) of the city as a space for interaction between strangers. This raises questions for urban design which are more social or cultural than technical, and concern the construction of the public realm as much as the design of parks and plazas; just as choices offered in consultation on urban design tend to be at a level of detail, so research such as Mozingo's could be misused to foreclose issues of the determination of the ambience of the city within which 'front-' or 'back-yard' experiences are desirable. Wilson questions recent writing which emphasises the need for safety as reconstructing paternalism, suggests that women's urban experience is more ambiguous than men's, and insists on women's rights to carnival, to the intensity and risk of urban living (Wilson, 1991: 9–10).

A cultural exclusion zone

Massey argues that the masculine public domain is the site for modernist literature and its characters, and Janet Wolff writes: 'The literature of modernity describes the experience of men. It is essentially a literature about transformations in the public world . . . What nearly all the accounts have in common is their concern with the public world of work, politics and city life' (Wolff, 1989: 141). At one level, this is explained by the extent to which professions, from politics to medicine and art, have discouraged women's admittance except in subordinate roles – men being, for example, government ministers, doctors and artists whilst women are secretaries, nurses and models. Massey writes: 'the mid-nineteenth century was a crucial [period] in the development of the notion of "the separation of spheres" and the confinement of women, ideologically if not for all women in practice, to the "private" sphere of the suburbs and the home' (Massey, 1993: 233). The scenes depicted in mainstream art from Impressionism to Cubism are spaces in which women, if present, are assigned the roles of actress, dancer and prostitute.[29]

FIGURE 17 The Guerrilla Girls, poster commissioned by Public Art Fund, New York (Photo: Timothy Karr, Public Art Fund)

Women are still excluded (except as images) from the museum presentation of art – the Guerrilla Girls ask 'Do women have to be naked to get into the Met. Museum?',[30] and the London-based group Fanny Adams applied the text '95% female-free' to a convenient white space on posters for the *Gravity and Grace* exhibition of sculpture at the Hayward Gallery – and from public art: most artists who achieve recognition in public art through either commissions or coverage in the literature are, like those in galleries, men,[31] the vast majority white, and the values conveyed by much public art have been seen as those of patriarchy.[32] How, for example, do women office workers and cleaners perceive Fernand Botero's globular *Venus* in Broadgate as they go into the building outside which it is sited? Is it a formal device, an organic (feminine) form of the architect's squiggle that signifies sculpture in contrast to the (masculine) geometry of the building? Or is it the earth-mother, related to men's notions of the primeval? And was Richard Serra's *Tilted Arc* a masculine gesture of dominance over a de-valued space, like a line on a sheet of paper? Suzi Gablik sees Richard Serra as setting up a win–lose, dominator–victim model of the artist's situation, and cites Barbara Rose in making a case that Serra's work is the product of a heroic and belligerent ego, proposing a world view which denigrates (feminine) empathy and relatedness to others (Gablik, 1991: 63).

Histories from which women are excluded are more than narratives of men's works, being the development of masculine attitudes to life, through role models and definitions of what achievements are worth recording. Griselda Pollock writes at the beginning of *Vision and Difference*: 'Is adding women to art history the same as producing feminist art history? . . . As early as 1971 Linda Nochlin warned us against getting into a no-win game by trying to name female Michelangelos' (Pollock, 1985: 1) and Gablik:

> Fitting into this myth of the patriarchal hero became the precondition
> for success under modernism for both men and women . . . Art as a closed
> and isolated system requiring nothing but itself to be itself derives from
> the objectifying metaphysics of science – the same dualistic model of
> subject–object cognition that became the prototype for Cartesian thinking
> in all other disciplines as well.
>
> (Gablik, 1991: 62)

Questioning the absence of women from art history can be only one point of departure for a re-writing of the terms of the discussion. A radical approach is informed by new work in geography, a subject which has until recently, according to Rose, 'been unusually reluctant to admit women and women's experience into its disciplinary imagination' (Rose, 1993: 70), which likens it to law, medicine, politics, architecture and art. Feminist writers in cultural geography, moving beyond – but not ignoring – questions of women's safety in the built environment (an area corresponding to that of creating access for women to art history), question the definition of place as something coherently bounded in space,[33] the primacy of vision over other senses, and 'transparency' of space,[34] as elements of a masculine view of the world. Massey argues that the vision is privileged precisely because it offers detachment (Massey, 1994: 232) and Sibley writes of the 'voyeuristic impulse, which Rose represents as the "masculine gaze", reifying and dehumanising the "other"' in relation to map making 'and other forms of distant geographical description' (Sibley, 1995: 185). How can this marginalisation of the feminine be addressed?

STRATEGIES OF RESISTANCE

Luce Irigaray, considering strategies of resistance to patriarchy's dominant cultural forms, writes, in *Thinking the Difference*:

> To anyone who cares about social justice today, I suggest putting up posters in all public places with beautiful pictures representing the mother–daughter couple – the couple that illustrates a very special relationship to nature and culture. . . . This can be done before any reform of language, which will be a much longer process.
>
> (Irigaray, 1994: 9)

It seems, at first, an almost quaintly avant-garde idea: to change the world through images, like David's *Marat*, Courbet's *Stone-breakers*, or the agit-prop trains of the Russian Revolution. That images influence behaviour in general is demonstrated by advertising, the images of which mirror back to us something of whom we are thought or required to be, but advertising images (including the mother and daughter who promote a brand of washing-up liquid said to be kind to hands) generally associate products with desires rather than social responsibility; whether the spell works when the implied gain is 'social justice' is uncertain.

Irigaray's proposal, if effected, would present at least two problems: one is the lack of a vocabulary of mother–daughter images in western culture, given the universal male child of Christian imagery and European kingship, so that mother–child has become mother–Son[35] – the problem Irigaray's proposal is intended to foreground;[36] the second is that the production of images involves codes and technologies of representation which raise issues for visual language equivalent in complexity to those encountered in any reform of verbal language – Craig Owens writes: 'What can be said about the visual arts in a patriarchal order that privileges vision over the other senses? Can we not expect them to be a domain of masculine privilege . . . of mastering through representation the "threat" posed by the female?' (Owens, 1983: 71). Massey, citing Irigaray several times in *Space, Place and Gender* claims that it is now a well-established argument that modernism sets up a way of seeing which is authoritative, privileged, and masculine (Massey, 1994: 232), noting Irigaray's statement that vision objectifies, hence controls, more than other senses.

Some images by women artists, such as Barbara Kruger's billboard text 'we don't need another hero', intervene critically in society[37] and Cindy Sherman intervenes in a cultural space to challenge the assumptions of how men look at representations of women, using the (perceived as transparent) medium of photography in her series of *Untitled Film Stills*. Massey, taking issue with Harvey's interpretation of the same images, clearly to his distaste, sees this as a way of disrupting the pleasures of the patriarchal visual field (Massey, 1993: 237–8) – a field in which 'woman' has been constructed as an object of men's scrutiny, the gaze disrupted when the female artist becomes an active subject constructing her own identities with the mobility associated with men (all on the move and on the make). The lack of a vocabulary of images of mothers and daughters, and a masculine dominance in codes of representation, are both problems which apply to imagery in any context – gallery, plaza or poster reproduction; but there is a third problem in that Irigaray proposes a public location for these images of resistance, in which they will compete for the spectator's attention with a variety of visual and tactile stimuli in a peopled streetscape. The street is not a neutral space, so this feminine imagery is to be located in a hitherto masculine realm.

The design of public space for women's ease might be one way into a more egalitarian public realm. But, it is more than a question of adding women, and women's experiences, to the urban scene; it is necessary to interrogate the way the city has been *conceptualised* as a foundation for a masculine public realm.

The absence of women in art history reflects the patriarchal society which produced that history as a set of categorisations of masterpieces, in the same way that general history affirmed masculine constructions of greatness, goodness and heroism; Pollock refers to a 'structural sexism' which 'contributes actively to the production and perpetuation of a gender hierarchy' (Pollock, 1985: 1). What we know of the world is, she maintains, a representation of the ideology of the social order within which that knowledge is learned. Irigaray argues that hierarchies operate in language to condition how we think, as when 'man', or 'mankind', is used as a universal term, and a plural referring to 999 women and one man remains in the masculine. In place of universalism, recognition of cultural difference and its implications has been proposed by both feminists and writers on post-colonialism,[38] and a redefinition of the terms of history (and relations to the body[39]) is as urgent for artists of colour as for women artists.

Some areas of current art practice by women take art as process not object, and involve participation rather than observation, for instance the 'new genre public art' (Lacy, 1995; Felshin, 1995) discussed in Chapter 8; similarly, male-dominated processes of planning and design might be changed through what

Gablik calls 'connectedness'. This implies challenging the Cartesian foundation of the dominant representations of space and its impact on urban design, and on modernism in architecture and art. Gablik, and more recently Lacy, foreground cases of interventionist art in the USA which challenge the foundation of the representation of space by attaching value to process rather than object (the process cannot be commodified and operates in representational spaces with identified groups of people), and are to a significant extent produced and critiqued by women.[40] 'New genre public art' has emerged following the impact of feminism and eco-feminism. Lacy sees a contributing factor in this development in the USA as a defence against the backlash against women in the 1980s, allied with a rising consciousness on the part of people of colour and concern about AIDS and environmental destruction. For Lacy, a combination of feminism, Marxism and ecology has created an urgency in re-visioning social values, and for Gablik, citing Adorno and Lyotard, it is also an effort to reconstruct a sense of meaning in life, despite Auschwitz and the post-modern rejection of meta-narratives.[41]

Reclamation

A society's value structure is evident in its classifications, such as gender and other mutable definitions including health and sanity, as well as its cultural production and language. Western society's image of the feminine has changed in history, through shifting perceptions of physiology, and through assigned roles and images which are gender-specific. Within feminism the body is seen in more than one way, either by denying its role as determinant of innate difference (because gender is cultural, whilst sex is biological), or by seeking to reclaim representations of women. Is it that, in a dualistic world, white, educated men think, and from their interiority gaze, on a world of extension in which women are objects in the field of view? Rose notes the role of medical science in establishing women and people of colour as 'other'[42] and continues: 'while the white male could transcend his embodiment by seeing his body as a simple container for the pure consciousness it held inside, this was not allowed for the female or the black' (Rose, 1993: 73). The masculine is, then, for the sake of an argument it has found useful, constructed as *rational, abstracting and analytic*; and the feminine as *intuitive, bodily and holistic*. If the masculine principle is divide and rule, the feminine is connectedness; if masculinity deals in representations of space inscribed on the (as if) value-free ground of the city, femininity occupies the representational spaces of lived experience and appropriation. These positions are theoretical polarities, and any characterisation of masculinity and femininity according to such simple aspects as mind–body, or competitive–collaborative is problematic in that other aspects, such as race, age, and class, determine other differentiations for both genders, and in that the oppositional presentation of pairs of traits is arguably itself a masculine approach, as if such complexities could be likened to the mechanism of a light switch which is either on or off.

But, if Irigaray's proposal for images of the mother–daughter couple is taken to indicate a 'possibility of difference', its challenge to the universal masculinity of the mother–Son might engender a challenge to the dominance of the representation of space; through a reclamation of the representational spaces of sociation and appropriation, in what Sennett calls the 'uses of disorder' (Sennett, 1996) in contrast to the bleakness of the purged city and the solitude of interiority, and through developing various means of empowerment, alternative ways of thinking about, practising and responding to planning, architecture, urban design and art might become possible.

3

THE MONUMENT

•

INTRODUCTION

The previous chapter considered issues of representation and the gendering of space. This chapter investigates the role of the monument, whether statue, memorial, or public sculpture, in maintaining the 'order' of western, industrialised societies.[1]

Monuments are produced within a dominant framework of values, as elements in the construction of a national history, just as such buildings as the Sydney Opera House contribute to a national cultural identity; they suppose at least a partial consensus of values, without which their narrative could not be recognized, although individual monuments may not retain their currency as particular figures fade in public memories, and individual buildings may be disliked. As a general category of cultural objects, however, monuments are familiar in the spaces of most cities, standing for a stability which conceals the internal contradictions of society and survives the day-to-day fluctuations of history. The majority in society is persuaded, by monuments amongst other civil institutions, to accept these contradictions, the monument becoming a device of social control less brutish and costly than armed force.

The use of culture as a means of preserving social order is stated as a general characteristic of bourgeois society in Herbert Marcuse's formulation of the 'affirmative character of culture'; it displaces value into an aesthetic domain, setting up a duality of art and life, allowing the impact of power or money on everyday life to be unquestioned, or at least less questioned. Whilst nineteenth-century monuments convey messages of empire and patriarchy, contemporary public art may be no less ideological in its content, regardless of its subject-matter. But are there cases of public sculpture which subvert the conventions of the monument, for example by a democratisation which celebrates 'ordinary' people, or by an inversion of its form, constituting a category of 'anti-monuments'?

ART IN THE STREETS

When contemporary art is sited in the street, two kinds of space collide: one is, as it were, set up by the 'autonomous' artwork around itself as an extension of art-space, which, like the modern gallery interior, is 'value free'; it sits comfortably within the conceptual spaces of city planning ('representations of space' in Lefebvre's terms) and equally value-free spaces of modernist architecture, so that there is a more or less easy relation of art to the design of the physical site. The other is a more informal and mutable kind of public space, the space around the bodies of city dwellers, termed by Lefebvre 'representational spaces'; this is always replete with values, personal associations, appropriations, exclusions and invitations, and the shared and disputed issues of the public realm, a set of overlaying spaces 'disordered' by users, and as such a psychological rather than physical space, which cannot be defined by map co-ordinates. The two spaces suggest different roles for 'public art': either public space creates wider access to the privileged aesthetic domain, but requires a level of cultural education if art is to be 'appreciated', just as the statue requires a recognition of its subject or type, recruiting more people to its liberal value-structure; or art, along with street theatre, street music and carnival, is a form of street life, a means to articulate the implicit values of a city when its users occupy the place of determining what the city is (Zukin, 1996: 43).

The first scenario, in which art in public spaces has an ideological aspect even when, as with abstract sculpture, it is presented as a purely aesthetic entity, its subject-matter the formal relations of shapes and volumes, sets public art within a history of the monument; it follows precedents in the commissioning of public sculpture between the 1870s and the 1920s,[2] which produced the statues and portrait busts which are now almost overlooked in the streets, squares and parks of most towns in Europe and the Americas, the allegorical figures adorning public buildings, triumphal archways and war memorials. The second has few permanent models, because it produces social processes rather than objects; one case might be the festivals organised by David for the Jacobins, another the agit-prop trains which ran the length of Russia after the Bolshevik revolution. Festivals were organised in Russia, too, sometimes including the unveiling of monuments; P. M. Kerzhentsev wrote, in 1918: 'art is breaking out of walls onto the streets' (cited in Tolstoy, Bibikova and Cooke, 1990: 53), and Nicolai Kolli's sculpture of the red wedge splitting a white block (a temporary work) was erected in Revolution Square for the first anniversary of the revolution.[3]

THE MONUMENT AND NATIONAL IDENTITY

Monuments stand in a complex relation to time: they state a past or its imitation, but are erected to impress contemporary publics with the relation to history of

FIGURE 18 These tourists are looking at a bronze Roman Emperor, but why? Not all monuments retain a currency of meaning, some being subsumed in the heritage fabric

those who hold power and the durability of that relation expressed in stone or bronze. One past out of many possible constructions is represented as *the* past, just as one concept of the city is represented as a dominant concept in its planning, as though history might be consistent whilst everything else, like our own lives, is mutable and temporary. This elevation to a realm of calmness and continuity is incidentally reflected in a description in *The Monument Guide* of an equestrian statue of *William III*, made by John Cheere in 1757 and sited in Petersfield as 'above the hustle of the market square' (Darke, 1991: 80), and in its inscription:

WHEN THE STATE WAS TOTTERING HE HELD IT FIRM

The equestrian pose, plinth and material, separate this king, in the constancy of a history in which the subject people are not intended to speak, from the noise and smells of the market; but resistance is never impossible and the statue lost its gilding when it was tarred and feathered during an election campaign. Darke continues: 'it has been called "ridiculous", but also "magnificent, heroic"; its counterpart, a bronze by Michael Rysbrack in Bristol, is without doubt a masterpiece' (Darke, 1991: 80).

The commonality, beyond that both are sited in public spaces, between public art and the monument is that both define and make visible the values of the public realm, and do so in a way which is far from neutral, never simply decorative. If the visual languages used seem diverse, the underlying model of intention may still have similarities: a mediation of history from the position of power is embodied in the monument, whilst the commissioning of public art is no less affirmative of given values or tastes, often dependent on state support, an issue taken further in the next chapter. John Willett, in his essay *Back to the Dream City*, writes of a continuity from the murals of the Nazarenes and Puvis de Chavannes, and the proposals of Watts and Manet to decorate the interiors of railway stations, to Tatlin's Monument for the Third International; these cases are, he maintains, steps in a development of the monument to fit modern secular democracy, and have a unifying source in the image of a New Jerusalem which 'goes on glittering away at the back of many minds' (Willett, 1984: 9). William Mitchell, who uses the terms 'public art' and 'monument' almost interchangeably in *Art and the Public Sphere*, sees another continuity, and asks if violence is central to the concept of the monument, noting that many memorials, monuments, triumphal arches, obelisks, columns, and statues refer to a past of conquest: 'From Ozymandias to Caesar to Napoleon to Hitler, public art has served as a kind of monumentalizing of violence' (Mitchell, 1992: 35) (see Figure 21).

Lefebvre's (and Massey's) characterisation of Picasso's art as violent in its distortion of the female body was noted in the previous chapter, but the task is accomplished in monumental art not through distortion but through sublimation; the dead heroes of conventional war memorials are relieved of their aggressive aspect and re-presented as a reflective and dutiful Everyman, embodying the required values of the humane state which nevertheless carries out its mission to rule. Similarly, statues of colonial administrators, generals and 'explorers' carry the white man's burden,[4] and the vanquished, too, became aestheticised. These are incidental cases of the way a public culture constructs a past, and may seem harmless when sufficiently distanced by time. According to Marcuse, however, not only does art in western culture generally perform a function of displacement of meaning to a conceptual realm, but this was a factor in the rise of fascism in the 1930s, the period during which his essay on the affirmative character of culture was written.

AFFIRMATIVE CULTURE

Marcuse's essay on 'The Affirmative Character of Culture' was first published in 1937. In brief, he states that in classical thought experience is separated into two realms: that of pure thought and leisure; and that of application through labour and utility. These are 'in a hierarchy of value whose nadir is functional acquaintance with the necessities of everyday life and whose zenith is philosophical knowledge' (Marcuse, [1937] 1972: 88). Thought's only objective is happiness. Marcuse sees in the division between beauty-leisure-peace, and labour-war-necessity the defeat of philosophy's original quest that practice should be informed by truth; this happens because the world of work and necessity is seen as insecure, so that if philosophical knowledge is concerned with a contentment which cannot be found in the material organisation of life, it must transcend it.[5]

A further development takes place in bourgeois society, characterised by its social contradictions and hence those of its thought, when the highest values of society cease to be the concern of, or appropriated as a profession by, the highest social strata, and a 'thesis of the universality and universal validity of culture' is devised by the new middle class; so, aristocracy is abolished to make the space of bourgeois democracy, which invents a culture of universal liberty, but people buy and sell labour in the *laissez-faire* competition bourgeois democracy also introduces, and 'the pure abstractness to which men are reduced in their social relations extends as well to intercourse with ideas' (ibid.: 93–4). To gloss some quite complex writing: after the abolition of the aristocratic privilege of pure thought, bourgeois society establishes a notion of universal cultural value whilst denying its applicability in a divisive system of labour relations. When alienation and commodification is structural in everyday life, the location of an idealised freedom must be somewhere else.

That 'elsewhere' is provided by the aesthetic dimension, through 'affirmative culture', which Marcuse defines as 'that culture of the bourgeois epoch which led in the course of its own development to the segregation from civilisation of the mental and spiritual world as an independent realm of value' (ibid.: 95). He argues that in bourgeois society, its value structure based in Reason but its claim of universal happiness an impossibility in face of the inequality set up by capitalist methods of production, 'the bourgeoisie could not give up the general character of its demand (that equality be extended to all men) without denouncing itself' (ibid.: 97); culture, a mental, abstract world, therefore takes on the claim to happiness, but is internalised, so that liberation becomes confined to the 'good, true and beautiful'. Bourgeois art subsumes the contradictions of life in fantasy: 'Only in the realm of ideal beauty, in art, was happiness permitted to be reproduced as a cultural value in the totality of social life' (ibid.: 117). Beauty in art is thereby compatible with social misery, and offers 'the consolation of a beautiful moment in an interminable chain of misfortune' (ibid.: 118). Art, then, depends on an internalisation of responses to life, a search for happiness diverted from actuality to an aesthetic dimension, as in the symbolist art of the 1880s–90s and the abstraction which followed it in the early twentieth century, and hence on an acceptance of the contradictions of society. When Huysmans wrote of Des Esseintes, that he withdrew from a vulgar world, he prefigured affirmative culture, whilst equally, perhaps, setting up the possibility for a self-contained 'system called' in which there might be a 'possibility of difference' (see Barthes, 1982). In a more pragmatic way, the model of affirmative culture could also be applied to Haussmann's re-planning of Paris in the 1860s. Haussmann has been described by Walter Benjamin as an 'artist of demolition'[6] and his work is seen by Susan Buck-Morss, writing on Benjamin's Arcades Project (Buck-Morss, 1995: 89), as a case of re-organising urban spaces and forms whilst retaining social contradictions and class antagonisms. At the centre of the boulevards is the Arc de Triomphe.

THE MONUMENT – HISTORY AND HEGEMONY

The expansion in the commissioning of monuments in the late nineteenth and early twentieth centuries, when the plunder of colonial wars was being assimilated to European museum culture (Coombes, 1994), is a statement of national identities which takes place in slightly different forms on both sides of the Atlantic.

The allegorical figures representing the *Four Continents* designed by Daniel Chester French[7] and completed in 1907, on the neo-classical US Custom House at Bowling Green, New York, are geography translated into female figures which signify cultural and political attitudes. They too form, in their similarities of language and intention but variance of detail, a system of difference, and in their time they stood for a web of relations between continents. *Asia* sits in contemplation, holding a lotus; skulls at the base of her throne remind us of the transience of life and supplicant figures suggest either the rejection of materialism or, in their emaciation, the inferiority of asceticism compared with American materialism. *Africa* (Figure 19) drowses bare-breasted, sleep a metaphor for darkness (and suggesting that this sleeping beauty is unaware of our gaze on her nakedness), resting one arm on a lion (primordial strength) and the other on the head of the sphinx (ancient mystery); she looks like a voluptuous but tired model escaped from a nineteenth-century *salon* painter's studio, playing, to invert the sentiment of a Barbara Kruger photo-collage, 'nature to our culture'. *Europe*, like *America* framing the central entrance to the building, is stately in Greek dress, holding a book and resting her arm on a globe – the word or the law, and the world – like a figure in judgement. Fragments of the Parthenon frieze are referenced on the side of the throne to afford a link to democracy, though a Roman imperial eagle has somehow alighted behind her, perhaps a Republican at heart like his American

FIGURE 19 US
Custom House,
New York – *Africa*

cousin. The one dynamic figure is *America* (Figure 20), for which read 'United
States', who seems to move forwards while a kneeling male figure turns a wheel
of fortune, or industry; she holds the torch of Liberty and a sheaf is placed on
her lap to signify abundance. A Native American is conscripted into a secondary
role behind her, visible only from certain angles, as if accepting a genocide

FIGURE 20 US
Custom House,
New York –
America

still, in 1907, within living memory. The building is now a museum of Native
American art.

The selectivity of the narrative and idealism of its visual language is obvious
when compared to, for example, documentary film, although that, too, makes
selections; other examples of nineteenth-century public sculpture, such as *Horace*

Greeley (1890) in City Hall Park, or *George Washington* (1883) at Federal Hall National Memorial, both by John Ward, utilise another visual language – the 'likenesses' of portraiture and naturalism – which seems more down to earth, but may be no less selective or artificial. Naturalism dispenses with the outward appearance of high culture – 'as if we might have known them' – though not the plinth or male domination of its subject-matter, but the transparency it offers disavows complexity through a focus on recognition rather than interpretation, which inhibits ideological readings but does not restrict ideological intention.

The associated meanings of a portrait bust may change in history. How, for example, does a black person in Brixton (site of what the media call 'race riots', where black people have died in police custody), whose ancestry might be traced to slaves on the sugar plantations of the West Indies, see a bronze bust of Victorian, white philanthropist *Sir Henry Tate*, who donated some of his fortune to the Tate Gallery and the library behind his bust, but made that fortune from the development of the sugar cube, sugar being a trade which depended on slavery?[8] Once this kind of change is perceived, the ideological basis of naturalism becomes accessible. The head of *Nelson Mandela* by Ian Walters outside the Festival Hall in London, though larger than life, is a modern form of naturalism; its reception has already shifted since its unveiling in 1985 by Oliver Tambo, with Mandela's release from prison and election as the first black President of South Africa.

Hegemony

The nineteenth-century development of monuments in the public realm, within a programme of public education and betterment undertaken by a state representing the industrial middle class, contrives to be a national story also told by the invention of traditions including tartans and morris dancing, as well as the opening of public collections of art. The contrivance is that the story offered is the only one offered, a process of persuasion in which the dominant class seems to 'naturally' inherit history; Gramsci termed the process of which this is an aspect 'hegemony'. Renate Holub gives a succinct definition:

> Hegemony is a concept that helps to explain, on the one hand, how state apparatuses, or political society – supported by and supporting a specific economic group – can coerce, via its institutions of law, police, army and prisons, the various strata of society into consenting to the *status quo*.
>
> (Holub, 1992: 6)

Frantz Fanon gives a similar description of how white society operates, which he contrasts with the more brutal application of law to black people under colonial administrations; he sees education, the handing down from father to son of moral reflexes, the loyalty of workers who are given rewards for long service, as 'aesthetic expressions of respect for the established order [which] serve to create . . . an atmosphere of submission' (Fanon, [1961] 1990: 29). But Holub goes on to add another dimension to hegemony:

> On the other hand, and more importantly, hegemony is a concept that helps us to understand . . . also how and where political society and, above all, civil society, with its institutions . . . contribute to the production of meaning and values which in turn produce, direct and maintain the 'spontaneous' consent of the various strata of society to that same status quo.
>
> (Holub, 1992: 6)[9]

In Chapter 5, this model is applied to property development, which serves the interests of a specific economic group and is frequently embellished with public art (Zukin, 1996: 45).

The proliferation of statues and memorials in the late nineteenth century sought to legitimise the recently acquired powers of the European or American nation states and wealth of their entrepreneurs; it subsumed social conflict within a myth of national identity, and (in Europe) personal grief for the deaths of colonial wars within a myth of sacrifice, supported by an established religion and commemorated in memorials which were sites of public remembrance, a process described by Bird as 'the state's desire to represent itself as the unifying authority' (Bird, 1988: 30). The transformation of memories into national memory replaces experience with a unifying abstraction to which the memories are co-opted, just as onto the remains of temples excavated during the Napoleonic campaigns in Egypt, for instance, was projected a notion of their past to which the ruins were then co-opted.[10] This does not deny a romantic reading of ruins as speaking of uncertainty against Enlightenment predictability, as in Shelley's poem *Ozymandias*, based on the colossus of Rameses II in his mortuary temple at Thebes (Figure 21) – a statement against autocracy as much as of the triumph of time; but it also allows Speer's 'theory of ruin value' – that buildings should be constructed to 'ruin well' in the thousand-year *reich* (al-Khalil, 1991: 38).

Legitimate memory

There is a history of architecture as persuasion through inevitability. The scale of fascist architecture and spectacle is one form of this, but others use history rather than might. For the emperor Hadrian, whose succession was dubious, history was used to invent a continuity into which to insert himself. His major work is the rebuilding of the Pantheon in Rome, on which he inscribed

M. Agrippa L. f. cos. III fecit

FIGURE 21 The feet of the colossus of *Rameses II* on which Shelley's poem *Ozymandias* is based

using not his own name but that of the builder of the old Pantheon. Hadrian resolved the tension between a need to establish himself through building and a need to keep a low profile by a revealing the city's 'essential' character through his monuments, confirming his legitimacy (Sennett, 1994: 94).

Something similar happened in the nineteenth century, but power was then located in an economic rather than military class; Bird writes of a legitimising which 'became the crucial operation for hegemonic structuring of civil society', and the 'recognition and celebration of hierarchical authority under the aegis of rituals and commemoration' (Bird, 1988: 30). The devices for this were statues and memorials, and rites of remembrance which replicate social hierarchies. Some of Michael Sandle's sculptures destabilise this tradition, as in *A Twentieth Century Memorial* (1971–8), made in context of demonstrations against the Vietnam War, in which a skeletal, blackened Mickey Mouse sits at a machine gun. *George and the Dragon* (1987) also subverts a heraldic device in an exhibition of masculine brutality, representing George as hard-working killer. Whilst Gramsci writes of the priest and pharmacist conveying the values of society to a peasant class in whose interests rejection would be healthier, Mickey Mouse stands for imposed social norms, and collapses the project in banality, while St George reveals a violence at the centre of those norms.

Conventional memorials present an idealised image of war which supposes a reverence for the nation rather than its conscripts – war is noble not bloody, death

does not really hurt. An example of heroic figuration is F. Derwent Wood's *Machine Gun Corps Memorial* at Hyde Park Corner – a naked David standing on a plinth, flanked by wreaths; it is hard to read from this the role of machine guns in turning a war fought on eighteenth-century infantry principles into mass slaughter in muddy ditches. A departure from such conventions, more radical in its time than it might now seem, is Charles Sergeant Jagger's nearby *Artillery Memorial* (Figure 22); Jagger served in Gallipoli and France and his monument accurately models every detail of clothing and equipment so that death, too, is stated in the pathos of the recumbent figure covered with a coat, a tin helmet on his chest. At the same time, the figures can be read as types rather than individuals, that is, as an aesthetic statement, framed by the latin text

UBIQUE

above a list of countries, expressing endless space as a metaphor for endless time rather than death's finality – because that finality is what crushes.

FIGURE 22 Charles Sargeant Jagger's *Artillery Memorial* in Hyde Park, London (detail)

THE LANGUAGE OF ALLEGORY

The mediation of history asserts its persuasion, then, through an aesthetic transformation. Despite the variety of forms, the common aspects are the primacy of a visual reading, based either in the idealising forms of classicism or the recognition of naturalism, and the communication of inevitability. Such a quest needs a coherent visual language, but to what extent is the language of the monument continuous through history? For Willett, it begins in 1789 with representations of Reason; yet he argues that the language used remained initially within an earlier, allegorical tradition, enabling, for example, 'Commerce to figure almost as a secular saint' (Willett, 1984: 8). Concepts arising from the new political consciousness of eighteenth-century bourgeois society such as *Liberty*, from colonialism such as *Navigation*, and from industrial expansion such as *Mechanics* or the *Telephone*, were given allegorical form, usually female.[11] Setting urban development within a classical framework conferred a degree of respectability; the entrepreneurial desire for expanding opportunities for manufacture and trade was supported by these images in public buildings which lent to industry the legitimisation of the state, and promoted the values of (men's) commerce by likening them visually to (feminine) images such as *Liberty*. The problem for sculptors was in applying allegorical language to modern rather than classical ideas, and the effort faltered in face of inventions which are not ideas but things; hence *Telephone* is represented merely as a reclining female figure holding a telephone,[12] and *Transportation* at Grand Central Station in New York,[13] unveiled in 1914, borrows the figure of Mercury presiding over a clock between Minerva and Hercules, although one of Mercury's roles was to transport the souls of the dead to the underworld, a reference presumably not intended for rail travellers.

Liberty, a political concept meaning the empowerment of the male property-owning class to carry out reforms of society, might seem more straightforward, but the difficulty of depicting it is seen in Delacroix's *Liberty Leading the People*, marking the July uprising of 1830. Delacroix both borrows and departs from allegorical conventions, using a realist language for the crowd, who are particularised in pose, dress and the kind of weapon they carry, whilst for Liberty reverting to a Greek type, as if a classical statue had been lent a tricolour and come to life. Marina Warner describes Liberty as 'in the classical costume of a goddess of victory, and her lemony chiton has slipped off both shoulders. Her breasts, struck by the light from the left, are small, firm, and conical, very much the admired shape of a Greek Aphrodite' (Warner, 1987: 271). The contrast between Liberty and the insurgents brings out the contradiction between realism and neo-classical allegory – for which Warner uses the phrase 'in a place of ideal difference' (Warner, 1987: 271) – perhaps between working-class 'freedom' (from oppression) and bourgeois 'Liberty' (to prosper). But could Liberty be presented in any other way? Delacroix's painting still met a hostile reaction when first exhibited in 1831, in part because the conventions of neo-classicism had slipped in Delacroix's handling, but more because, with a restored political order, a red bonnet had become a subversive emblem.

Goya's *The Third of May 1808*, painted in 1814, is more consistent in its language, depicting the victims of execution with emotive realism and posing the execution squad so that their faces are unseen – a pictorial device to dehumanise them used also in *One Can't Look* (Disasters of War 26). For a European viewer, the white-shirted victim with arms raised takes on the identity of a Christ-like martyr, the firing squad of Roman soldiers, or fascists. The *Cable Street Mural* (Figure 23), commemorating opposition to an attempted fascist march through London's East End in 1936, uses the same device to evoke sympathy with resisters to fascism:[14] the demonstrators are shown full-face, but the policemen are painted either in back-view or with helmets pulled down. The *Cable Street Mural* establishes its cast as heroes and villains, as crudely as a spaghetti western but using pictorial devices from art history.

Victory and liberty

National memory is still constructed through public monuments, not only in the West. Saddam Husain's *Victory Arch* in Baghdad (which he designed himself in anticipation of a victory, which was never recognised, in the Iran–Iraq war) consists of two pairs of giant arms scaled up from casts of his own (cast in a foundry in Basingstoke) holding swords cast (in Iraq) from the weapons of dead soldiers, placed at each end of a parade ground. At their bases, five thousand Iranian tin hats topple in a heap. Samir al-Khalil locates the Arch in the context of Iraq's utilisation of heritage culture and replication of archaic sites within the modern state as backdrops for images of the President, suggesting a continuity in which Saddam Husain is the most recent link to a history which includes Babylon;[15] he contrasts the realism of the arms with the falsity of the message of victory and notes the proliferation of monuments, mosques and memorials of huge size under the Ba'athist regime. The Gulf War now colours western perception of *Victory Arch*, but is its lack of authenticity different from that of, say, the *Statue of Liberty*, a monument also inserted into public memory for ideological reasons?

The *Statue of Liberty* has a longer history than *Victory Arch*, and being visible from Ellis Island where millions of refugees awaited entry to 'the land of the free', is a symbol of the opportunity to escape poverty. As an idea, its history begins with the French liberal jurist Edouard de Laboulaye's proposal that France should present a gift to the United States on the centenary of its independence – an oblique comment on French politics under Napoleon III. As an image, it derives from precedents such as Egyptian statues, and the more immediate source of neo-classical public sculpture; Frédéric-Auguste Bartholdi, who visited Egypt with the orientalist painter Gérome, was commissioned by Laboulaye in 1865 to make a portrait-bust, and, at the time, was in process of conceiving a grand monument for the entrance to the Suez Canal, which he called *Egypt Bringing the Light to Asia*.

FIGURE 23   The face of a policeman covered by his helmet, *Cable Street Mural* – the white drips are the result of paint-bombing by right-wing extremists

It was this image which became the basis for Liberty when the project was taken up following the restoration of the French republic; Bartholdi went to the United States in 1871 and began adapting his idea for the entrance of New York harbour, exhibiting the arm with beacon at the Philadelphia World's Fair in 1876, and the head in Paris in 1878; the gift was funded by public subscriptions raised in France, formally accepted by Congress in 1877, and unveiled in 1886 after subscriptions raised in the United States paid for the pedestal. Bartholdi's Liberty is very different from Delacroix's – lacking a red cap as Laboulaye observed. Warner echoes this reading: 'Liberty is no longer La Liberté, but was identified from the start with an American ideal of democracy, now represented as an American gift to the world' (Warner, 1987: 7). Its dominant reception, through popular prints and in the literature of shipping companies, is as a mother of exiles; but there is a contradiction – the first laws restricting immigration into the United States, banning convicts, lunatics and Chinese labourers, were passed as the statue was being commissioned.

There is a commonality between Bartholdi's *Statue of Liberty* and Saddam Husain's *Victory Arch* in the way each signifies an inevitability. Liberty has come to stand for democracy, and America as its purveyor, so that opposition to American interests threatens democracy, which in turn is, like high art, a defining aspect of western civilisation. Victory Arch seeks to establish the Ba'athist regime as inheriting a history going back to the dawn of urban settlement in Mesopotamia, so that opposition to it is similarly an assault on the timeless values of civilisation. If the meaning of the *Statue of Liberty* embodies contradictions, its appearance undergoes mediation in public appreciation. Warner describes it in terms of 'a thunderbolt judge of stern unrelenting character' (Warner, 1985: 8–9) and contrasts this with the soft-focus photography used in souvenir illustrations. At the same time, demonstrators in Tiananmen Square made a version of the *Statue* to signify their aspirations to freedom in 1988 (Mitchell, 1992: 29–31, 37); just over a century before, in 1885, following the new restrictions on immigration, a Chinese immigrant wrote that to him the appeal for a subscription for the statue's base was an insult.[16]

FRAMEWORKS

Frameworks of discussion on monuments differ according to professional perspectives. *The Monument Guide* (Darke, 1991) gives a non-critical description of monuments as items in a geographical rather than cultural landscape – 'an open air treasure house' or 'topographical gallery' (Darke, 1991: 10). Marina Warner (Warner 1985), reading monuments through cultural history, sees those which have continued to receive wide public interest, despite the loss of their original purpose, as cultural signifiers for a city. She writes of the *Statue of Liberty* that it shares characteristics with the Eiffel Tower, that both are first monuments with little other function – 'Stripped of use and service, they resist obsolescence. They are in the first place expressions of identity' (Warner, 1985: 6). The idea that monuments stand for cultural identity is extended by Donald Horne to include everyday items: he links, for example, the Ramblas with Barcelona, and argues that 'a meat pie eaten in the street is part of the "language" of Melbourne' (Horne, 1986: 42–3). Signifiers of place are not always associated with wealth – Wigan Pier signifies the recession of the 1930s but has become a tourist attraction. Sociologist John Urry, who coined the term 'tourist gaze', suggests that unlikely places become centres of heritage-based tourism (Urry, 1990: 105) and that such survivals are relieved of meaning as a memory of industrial exploitation, and re-presented in a sanitised form (Urry, 1995: 160); he notes Robert Hewison's argument (Hewison, 1987) that heritage culture constructs an illusory past which distracts attention from present conflicts, though he defends the visitors' centre at Wigan Pier as preserving local memories, some of which are oppositional. Local authority officers, museum curators or tourist organisations aiming to develop heritage industries, however, tend to select safe aspects of history for public consumption; John Molyneux, a social historian, states of their efforts: 'it is much more important to tell a pleasing story, to give tourist and visitor what they want, than it is to be true to history' (Molyneux, 1995: 16). Social and economic histor-ians such as Molyneux, on the other hand, interested in the structures of power and money in industrial societies, and to expose contradictions in how a society represents itself, have tended to see monuments and heritage culture as devices for social control. Molyneux continues: 'Nine times out of ten the interpretation of history deemed non-controversial for heritage purposes is the right-wing view, the establishment view' (Molyneux, 1995: 18). Just as monuments construct hege-mony, so heritage sites construct a past which conveniently fits civic aspirations and serves social stability.

Two cases of art which illustrate different methods of constructing place are the *Miners' Memorial* at Frostburg, Maryland, also known as *Prospect V–III*, and *Cincinnati Gateway*, both by Andrew Leicester. The *Miners' Memorial*, of 1982, combines art, local history and community participation to commemorate a marginalised past. Old mining artefacts donated by members of the community are integrated in a series of passages and rooms through which the spectator progresses as in a journey through life from infancy, represented by a coal-cart as cradle, to death as the black shaft, in which a bed is sculpted from coal, and on the bed a figure.

Cincinnati Gateway (1988) in Sawyer Park on the Ohio river, was a more ambitious project, with a larger budget and more constraints in terms of the image of the city desired by those who exercised influence over its affairs.[17] If the public for the *Miners' Memorial* can be defined as the ex-mining community, it is more difficult to say for whom *Cincinnati Gateway* is intended. Described by John Beardsley as 'the most successful use to date of locally relevant – even reverent – imagery' (Beardsley, 1989: 144), it still seems a representation of a 'decorative past' which commodifies place in the same way as heritage culture. A walkway on an embankment in which is set an image of the river, a brick wall representing a canal lock, a bridge referencing the structure designed by John Roebling which spans the river nearby and is a forerunner of the Brooklyn Bridge, and sculpted images of fish, fossils, Native American artefacts, riverboat chimneys and winged pigs are some of its elements. Leicester stated: 'The sculpture's emphasis was supposed to be about Cincinnati over the past two hundred years . . . I expanded that narrow view of history by recalling the real social and cultural roots of the riverfront' (Doss, 1995: 208), which led him to include pigs signifying 'porkopolis': the city as a centre of pig meat production. Some voices sought to exclude the pigs in a controversy inflamed by local political rivalries; the *Cincinnati Enquirer* published a front-page article against the project in which the word 'pig' was used over twenty times (Doss, 1995: 225). Although the work was completed, and has received wide and generally favourable publicity, Leicester's representation of a city through a series of motifs created more than one public: one supportive, though perhaps for reasons also connected with objections to arts censorship; one antagonistic, again perhaps with its own agenda. Overlooked by both was the co-option as decorative motif of Native American imagery.

One of the arguments against the imposition of a contrived sense of place is that it aestheticises the city; another that it is a closed history which affirms the structures of power, taking the time of history and, as it were, spreading it out in a static present, so that projects such as *Cincinnati Gateway* are spatial representations of time. Ernesto Laclau has argued that whilst space is stasis, time is a process of dislocation which opens the way for political change.[18] Massey interprets the theory:

> Laclau, for whom the contrast between what he labels temporal and what he calls spatial is key to his whole argument, uses a highly complex version of this definition. For him, notions of time and space are related to contrasting methods of understanding social systems.
>
> (Massey, 1994: 251–2)

Any self-contained system, or causal structure in which change comes from inside and is thus 'unsurprising', is for Laclau spatial, whilst time indicates dislocation, disruption of the chain of cause and effect – 'The spatial, because it lacks dislocation, is devoid of the possibility of politics' (Massey, 1994: 252). It is not important, in this discussion, that the terms 'temporal' and 'spatial' are used, as much as that the distinction between the seamlessness ascribed to space and the dislocation ascribed to time from which change, or freedom from the replication of past patterns, is possible. The idea of a self-contained system could be mapped onto contemporary art, and the model of space and time onto art as object in space or process in time. So, elements of contesting pasts are de-activated in a continuous, seamless space, offering only a passive reception which, because it cannot admit change, is what Laclau, for whom 'Freedom is the absence of deter-mination' (Massey, 1994: 253), terms 'closure'. Massey is critical of Laclau for privileging time as the ground of being, and much of her own writing has sought a rehabilitation of space, but the idea of space as a closed system describes cosmetic notions of a 'spirit of place' dependent on the primacy of visual representation.

There are, then, three frameworks through which to interrogate the notion of the monument: following Bird and Molyneux, as the imposition of an ideology; following Darke and Warner, as landmarks or signifiers of place, to the consump-tion of which Urry and Hewison apply a critique; or, perhaps a third possibility is that the monument can be democratised.

The first two categories overlap. *Nelson's Column*, for example, was, when erected in 1842 an image of English naval power in the aristocratic neo-classical style; since then it has become one of London's postcard views, no doubt as popular with French tourists as any others. The only thing it has never been is a statement for a defined community, since the habitations which once filled the space over which it now towers were cleared to build the white, utopian vista of Trafalgar Square. The *Albert Memorial* (1862–72), designed by Sir Giles Gilbert Scott, was also part of the construction of a history, presenting Albert as a progressive consort who favoured industry and design – he holds the catalogue of the Great Exhibition of 1851 – and inherits the mantle of the chivalric past through the referencing of medieval Eleanor crosses; and it is a convenient landmark for people walking through the park. Tancred Borenius described it in 1926 as 'an episode in the landscape of Kensington Gardens – especially in the mellow light of the late summer evenings, when the masses of foliage are full and rounded' (cited in Darke, 1991: 60). But this fading into the urban landscape itself contributes to a de-politicisation of Albert.

RESISTANCES AND THE DEMOCRATISATION
OF THE MONUMENT

Challenges to the classical language of allegory began in the nineteenth century and include Rodin's *The Thinker*, in a version enlarged from the original and sited outside the Panthéon in Paris in 1906, which uses a realist language to express a sympathy for the emancipation of the worker, although in keeping with the cultural climate of its time it assumes a universal masculinity, and occupies a plinth. Rodin described it as 'the fertile thought of those humble people of the soil who are nevertheless producers of powerful energies'[19] and saw it as a precedent for a monument to French workers.[20] Its visual language supports this, given the association in French nineteenth-century art of realism and radical politics; Albert Elsen sees it as a 'culmination of his efforts on behalf of the workers' (Elsen, 1985: 107), noting the frequency of strikes and growth of syndicalism in France at the time.

If, then, realism was the art of radical politics in France in the mid-nineteenth century, are there cases of recent art in public spaces which attempt a comparable programme? In a way, Raymond Mason attempts this in *Forward* (Figure 24) in Centenary Square, Birmingham; the sculpture depicts the 'people of Birmingham' walking in procession towards the future (aligned in the direction of the new convention centre representing new investment). Yet Mason's figures are stereotypes, for whom no models in everyday life could have existed, with the same difficulties of reception as the monuments of socialist realism, many of which were being craned out of squares in eastern Europe as Mason's work was installed.

Statues to local people, unlike stereotypes, may retain meaning in local memories – for example, Henry Blogg, a lifeboatman commemorated in Cromer, Norfolk (Darke, 1991: 197), or *Snooks the Dog*[21] in bronze on the seafront at Aldeburgh in Suffolk (Darke, 1991: 195–6). A more direct immortalisation is Jim the station dog at Slough railway station, who is stuffed, in a glass case, being simply himself dead rather than a representation, though, of course, framed by the glass case with its museum associations and re-coded as memorial. Ticket collectors and porters have not received a similar treatment.

Kevin Atherton's bronze figures at Brixton railway station, *Platforms Piece* (1985–6) are life-casts (regarded as a single artwork) of three people who regularly used the station at the time, set without plinths on the platform. Atherton, who regards the piece as significant in his own development and that of public art, selected two black models and one white – Peter Lloyd, Joy Battick and Karin Heistermann:

> I have used the station platforms . . . as plinths for the three figures, thus pitching them into the same reality as the railway passenger. The work is about speculation – the speculation about other travellers that passes through one's mind when isolated and waiting for a train. By using younger people as models, I have also tried to capture the vibrant optimism of youth. . . .
>
> (British Rail, 1986)

It was claimed that local people had given the station a new identity, and that 'at Brixton, which is not the wealthiest part of the Home Counties, there has been no vandalism against Kevin Atherton's amusing bronzes' (Hughes, 1988) – no longer a possible claim since right-wing extremists have daubed the face of

FIGURE 24 Raymond Mason's *Forward* in Centenary Square, Birmingham

FIGURE 25 Kevin
Atherton, *Platforms
Piece,* Brixton
Station, London
(Photo: Kevin
Atherton)

FIGURE 26 Kevin
Atherton, *Platforms
Piece* (detail)

one sculpture with white paint, and never a very helpful one in linking a lack of wealth rather than lack of ownership with vandalism. The spaces of the station remain covered in graffiti and its interior reeks of urine.[22]

Anti-monuments

Other kinds of art in public spaces are more overtly challenging, such as Jenny Holtzer's texts on electronic message boards, or Krzysztof Wodiczko's projections onto monuments; and some groups, notably The Power of Place, a multi-disciplinary team working with minority communities in Los Angeles to recover their memories of place (Hayden, 1995), seek to build continuing and empowering engagement with identified communities. The issues are brought into focus when artists are commissioned to make monuments for the Holocaust, a 'buried' history. The *Harburg Monument Against Fascism* (1986) by Jochem and Esther Shalev-Gertz is sited in a shopping area half an hour from the centre of Hamburg; its form is a 12-metre lead-encased aluminium column initially raised above, then periodically sunk into, a pit. The artists set a text beside it:

> We invite the citizens of Harburg and visitors to the town, to add their names here to ours. In doing so we commit ourselves to remain vigilant. As more and more names cover this . . . column, it will gradually be lowered into the ground. One day it will have disappeared completely.
>
> (cited in Young, 1992: 56)

Many kinds of graffiti were added to the column, exposing prejudices against 'guest-workers' reminiscent of anti-semitism, the controversy around the monument accepted by the artists as part of its impact. The column has been described as an anti-monument, which refuses the forgetfulness of conventional monuments which do society's remembering for it (Young, 1992: 55).

Theodore Adorno argued that it is impossible to write *lyric* poetry after Auschwitz, but the argument has a general application; whilst the monument as a device of hegemony establishes a national history, so it may also bury a national memory. It seems interesting that some of the Holocaust or anti-fascist monuments in Germany, such as the column in Harburg, finally submerged in 1990, and Horst Hoheisel's inversion of the *Aschrott-Brunnen Monument* in Kassel, are, literally, buried, in reference to the invisibility of a particular history. The original fountain in Kassel was funded by a Jewish entrepreneur in 1908, and was removed by the Nazis in 1939. The artist describes his idea as a mirror image of the old structure, sunk beneath its location, 'in order to rescue the history of this place as a wound and as an open question' (cited by Young, 1992: 70) so that history is not repeated. It presents a monument-shaped hole into which water runs from a surrounding pool; from a distance only the sound of water indicates its presence.

FIGURE 27 Maya
Lin's *Vietnam
Veterans Memorial*,
Washington, DC

The history represented by statues is a closure inhibiting the imagining of alternative futures by denying the possibility of alternative pasts; but if this monument is an opening in society's received structure of values, dislocating the assumptions of an 'official' history, it is an act of resistance. It is in such a dislocation of the conventional reading of monuments that Wodiczko's projections work, abolishing the familiar and reminding us, for instance, by projecting onto it the image of a missile, that *Nelson's Column* is a monument to war. Wodiczko writes: 'The intervention lasts longer than the work itself . . . the power of the projection can be better understood when the projectors have been switched off. Something has been broken' (Freshman, 1992: 105).

Maya Lin's *Vietnam Veterans Memorial* (Figure 27) in Washington, DC, of 1982, consists of a shallow V-shaped 'cut' into the green of the Mall, one side pointing to the *Lincoln Memorial*, the other to the *Washington Monument*; it refers to a war the history of which is, in the light of protest and the social invisibility of the returning veterans, problematic. On its polished, black stone surface are inscribed the names of the 58,000 Americans dead or missing in action between 1959 and 1975, arranged so that the first and last are adjacent at the centre. Charles Griswald has described the memorial as open and closed, like a book (Griswald, 1992: 104) and cites Maya Lin as describing it in terms of a 'gash in the earth';[23] the scar it seeks to heal is perhaps the rent in American society caused

FIGURE 28 The
East Coast Memorial,
Battery Park City

by protest against the war. Beardsley writes that the debate around the monument
was a catharsis and that its success is in terms of 'easing trauma into memory'
(Beardsley, 1989: 124–5). Whether, in the end, the *Vietnam Veterans Monument* is
public art, architecture, or a monument, is not an interesting question; what seems
more to the point is whether it, or public art in general, affirms or interrogates

the structures of power in society which bring about wars, whether contradictions are addressed or buried, whether it constructs, in Laclau's terms, an open time or closed space. The *Vietnam Veterans Memorial* is not an anti-monument like the fountain in Kassel, and there are precedents for its black stone, including the base of the *East Coast Memorial* (Figure 28) in Battery Park, New York; but it avoids the idealising language of allegory and the replication of social hierarchies. It has created a public, being visited constantly, even on days that are cold and wet, by those who look for or make a rubbing of a name, pause, leave a flower, or simply look into the stone's mirror surface, or cry.

4

THE CONTRADICTIONS OF PUBLIC ART

•

INTRODUCTION

The previous three chapters set a context within which to theorise and criticise public art – by examining the conceptualisation of the city, by enquiring into the representation and gendering of urban space, and by linking the monument to hegemony in bourgeois society. This chapter considers the literature of public art – after asking if all art is 'public', and interrogating the notion of 'art-and-architecture' – and differentiates the literature of advocacy, which has supported public art as a practice in a critical vacuum, from that which relates it to social issues and might act to re-focus the practice in ways which are more likely to engender social benefits or lead to social change. It takes this approach – of seeing public art through its literature – because other sources already give various kinds of survey,[1] and because the literature is an informal demarcation of the territory, with many overlaps and disputes but more consensual than a definition imposed by one author.

The map of public art is difficult to delineate, and contested, but its polarities could be stated as, on one hand, a contemporary equivalent of the nineteenth-century monument, a practice which accepts social and artistic conventions, its contradictions concealed by relocation to an art space outside the gallery or museum and by the lack of documentation of its reception;[2] and an emerging practice of art as activism and engagement. These categories suggest, respectively, aesthetic objects and ideologically aware processes, or statements of closure (in Laclau's terms) and strategies for intervention. Central to the development of new, more theorised practices of public art is the recognition that there is no 'general public' (only a diversity of specific publics), and the redefinition of its location as the public realm, rather than a physical site assumed to grant access to an undefined public. Many cases of public art are somewhere between the polarities, partially aware of their political and social implications, yet adhering to conventional vocabularies of form and governed by the need of artists and arts managers to make a living.

PROBLEMATISING PUBLIC ART

After around thirty years of a discrete practice of public art which has spread to most industrialised countries, the time is ripe for a retrospect. The lack of a critique which includes insights from outside the institutions of art, for example through urban sociology, geography and critical theory, or through the responses of publics in whose spaces public art is sited, is, as stated in the Introduction to this book, an impoverishment of the practice, and a reason public art often fails to create a public. If that was simply a case of art being disliked or ignored, as in a bad review of a gallery show or failure of an art auction to induce adequate bids, it would concern only those with vested interests, but public art inevitably operates in the public realm and a lack of critical engagement with the construction of that realm leads by default to affirmation of the dominant ideology.

Descriptions of public art are generally not kind and reflect its marginality. It has been called 'a special kind of socio-aesthetic pudding' (Willett, 1984: 11), and 'an oxymoron . . . resolved in favour of banality' (Brighton, 1993b: 43), which indicates that there is a problem in using the term 'public art', and perhaps it is no longer possible to do so for anything other than a co-option of art to public policy through public funding. But, another writer states that artists 'working in the public interest address a wide range of human concerns' (Raven, 1993: 4), though this does necessarily mean that art as a social practice occupies only non-gallery sites,[3] and some art outside the gallery may still be more in the interests of the artist or curator than the (often unspecified) public. John Willett writes of 'art and society':

> over the past hundred years or so these words have become inseparables,
> till today their coupling produces an automatic yawn. They are as much
> of a pair as oil and vinegar, or chalk and cheese; like them they owe their
> association in our minds largely to the fact that they don't mix.
>
> (Willett, 1967: 1)

The terms 'art' and 'public' fit no more easily together in the twentieth century, and a definition of 'public art' is fraught with the contradiction that whilst modernist art has occupied the hermetic space of the white-walled gallery, art forms more closely linked to areas of everyday life, such as 'community arts' or 'outsider art', have been marginalised by the art establishment[4] as lacking 'aesthetic quality'. But, a sculpture in a plaza is not made accessible by its site as such, and any work of art in a public collection might be described as 'public',[5] so that the issue becomes not 'public art' but 'the reception of art by publics'. That reception can be manipulated.

Institutions such as national galleries and museums of modern art have since 1945 been central to the formation of dominant cultures. In Federal Germany it

FIGURE 29 Isamu
Noguchi, *Red Cube*,
New York

became almost a badge of anti-fascism to commission modernist art, and a work
as opaque as a Jackson Pollock drip painting – which could be said to have
nothing to say[6] – was used in international touring exhibitions during the Cold
War to promote the intellectual (not economic) freedom of the western artist
compared to the supposed un-freedom of the artist in socialist countries (where
they enjoyed considerable state support through artists' unions), its lack of
accessible content being part of its hegemonic function,[7] whilst its individualism
and 'risk' mirrored the ideology of the political establishment of the USA at the
time (Guilbaut, 1983). Modern art thus became identified with 'the West', which
also means with the 'interests of capital', today a complex web of multi-national

corporate entities. Much public sculpture in the UK and USA today is selected within the canon of modernist taste and is thus similarly complicit in its agenda – what do Noguchi's *Red Cube* (Figure 29) outside a bank head office in Manhattan, or Serra's *Fulcrum* (Figure 30) at the entry to London's Broadgate development, say, apart from that the bank and property developer are successful enough to be patrons of art (as of architecture) and therefore part of the continuing construction of liberal civilisation? And why is minimalist sculpture commissioned for corporate sites in Tokyo and Seoul if not as a signification of Japanese and Korean desire to subscribe to liberal *civitas*? Perhaps there is also something of this desire in the commissioning of 'major' works of public art by civic authorities, such as Antony Gormley's aircraft-like *Angel* for Gateshead.

FIGURE 30
Richard Serra,
Fulcrum, Broadgate,
London (detail)

Art and architecture?

Institutional public art is often legitimised through an appeal to a supposed past union of art and architecture, in another notion of liberal *civitas*. The Parthenon frieze, the sculptures of Chartres Cathedral and Egyptian temples, have been called on, as if in some kind of continuity, to support this (Jencks, 1984); Deanna Petherbridge, typifying the genre, writes:

> The coming together of artists and architects has always marked a period
> of creative vitality in Western cultural history. Such collaboration was
> central to the Italian Renaissance and left its mark on the fabric of cities
> as richly as it endowed palaces and churches.
>
> (Petherbridge, 1987: 2)

One, quite basic, problem with this is that the terms 'artist' and 'architect' are
specific to modern, Western culture and outside the conceptual categories of earlier
and other societies — there is no 'always' about this rich endowment and the
proposed re-marriage seems a kind of naturalisation of what is, in fact, a modern
professional commonality from which neither art nor architecture, despite a few
squabbles, has been divorced; another that the statement concerns only dominant
or privileged forms of culture, ignoring vernaculars and sub-cultures. But the
advantage of the past for those who propose a union of art-and-architecture is that
it is a ground on which to project a compensatory fantasy of the present which
abolishes conflict; such universals elide differences of race, gender, and class, whilst
tending to be, as Doreen Massey writes: 'white, male, western, heterosexual, what
have you' (Massey, 1994: 240). Heritage culture constantly uses this opportunity
to fantasise a past, and so, it seems, does art-and-architecture.

Whilst feminist cultural criticism situates the subject and deconstructs universals, the term 'art-and-architecture' represents a universalizing fiction. The
Parthenon frieze analogy, like the Florentine Renaissance, was a common theme
in conferences on art-and-architecture through the 1980s, but is, in effect, a means
of co-opting art to bourgeois notions of urban 'beautification'.

Sandy Nairne wrote in 1991, as Director of Visual Arts at the Arts Council:
'There is a natural bond between good contemporary art and craft and good contemporary architecture' (Arts Council, 1991: 9), but an alternative reading of the
situation by Andrew Brighton is that, forgetting the arcadian fantasies, when artists
and architects collaborate, each challenges the other's professional ideology
(Brighton, 1993a). Whilst collaborations were generally proposed to 'humanise'
architecture, Brighton is less polite, observing: 'the almost obscene spectacle of an
attempt to create siamese twins out of two corpses' which produces 'decorative kitsch
and authoritatively, banal buildings [sic]' (Brighton, 1993a). But does Brighton
mean that artists and architects have opposing ideologies, their meeting provoking
a mutual reassessment? If not, if it is only their working methods and terminologies
which differ, why should their conjunction change anything? He seems ambivalent;
on one hand he accepts the similarities: 'The notion of great artists or architects
serves to bolster the status of current practitioners; they are the descendants of gods'
(Brighton, 1993a), but on the other claims that 'for artists to make drawings from
which others can construct something, a drawing devoid of the rhetoric of self-
expression, is for some a radical act', though it was a common practice for Henry

Moore, and much recent art has rejected the craft notion of the hand-made by using readymades or industrial or electronic technologies. A reassertion of the hand-made, or craft object, as a step towards a post-industrial culture could be seen as radical.

A stronger case might be made for the commonalities between art and architecture as professional activities: both enjoy the notion of individual authorship; the separateness of art and architecture from society enables both to develop mystiques which protect from scrutiny and critiques which are self-referential; and both are immersed in modernist notions of quality expressed in terms such as 'innovation and excellence'. The result of collaboration, then, may be a strengthening for artists and architects of their existing view of space as value-free, society as 'other', and their professions as revolving around individualism, innovation and the interiority of the studio.[8]

But perhaps an adversarial model, in which artists and architects confront each other like the romantics and neo-classicists in the *Salon*, is not what Brighton intended. He writes: 'it is troubling to most architects to work on the assumption that a judgement of what is functional and what serves utility is as ideologically contingent as is an aesthetic judgement' (Brighton, 1993a); but what he does not establish is how artists, rather than critical theorists or post-modern geographers, will contribute to the awakening.

In a second, more considered text, Brighton develops his idea in terms of Eco's *Function and Sign: The Semiotics of Architecture* – 'in Eco's terms, the architecture of the Broadgate Development is mass culture. Its messages are clear and familiar. Its eclecticism slides easily into the architectural mix of the city' – whilst (modernist) art maintains its status as art by not being mass culture (Brighton, 1993b: 46–7); he assumes that art is to a degree at least a critical activity, though it could be suggested that for modernist art dissent is usually an argument about style. Brighton then argues that 'for the developer, the disjuncture between art and audience has a function; it has a general connotation of cultural authority and with it a concern for "quality"', whilst Serra's *Fulcrum* 'puts the habitual into question' (Brighton, 1993b: 48). He also refers to the terminal at Stanstead Airport, but as a site in which art is a nuisance: 'airport requirements, kiosks, eating and sitting areas, security barriers and so on, are like chronic coughing in a concert. But the aberrant icing on the cake, like someone whistling along with the orchestra, is the addition of art' (ibid.: 48). His provisional conclusion is that public art, if serious, is at variance with its context, whilst in architecture that denotes itself there is no need for decoration. This seems to leave public art only an activist route if it is to avoid banality.

Suzi Gablik, writing in *Mapping the Terrain*, cites an interview for *Flash Art* in which Christo asserts 'The work of art is a scream of freedom' (Gablik, 1995: 78); Gablik links this to Serra's response 'awash' in 'bad modernism' to the controversy over *Tilted Arc* (Figure 31), continuing:

FIGURE 31
Richard Serra,
Tilted Arc, New
York
What the *Tilted Arc* controversy forces us to consider is whether art that is centred on notions of pure freedom and radical autonomy, and subsequently inserted into the public sphere without regard for the relationship it has to other people, to the community, or *any* consideration except the pursuit of art, can contribute to the common good.

(Gablik, 1995: 79)

There is, then, a polarity: of art (and architecture) which is part of the dominant conceptualisation of culture, space and the city, in which public art is an element in the inscription of the city, as if on a value-free ground; and art with a social purpose, which seeks to re-define the values and open up the possibilities of urban living; Rosalyn Deutsche writes: 'Given the proliferation of pseudo- and private public spaces, how can public art counter the functions of its "public" sites in constructing the dominant city?' (Deutsche, 1991a: 167) to which the answer is by problematising the notion of 'public' as an ideological construct and dumping the baggage of civic beautification; she draws attention to the colonisation of public space and the city by the agencies of redevelopment, and states: 'Public art shares this plight' (Deutsche, 1991a: 168), adding that today's complexities require a new framework of analysis. That 'framework' is beginning to appear in debates on urbanism rather than art.[9] Art-and-architecture, meanwhile, acts as a framework for a complicit and institutionalised (and male-dominated) public art.

THE LITERATURE OF PUBLIC ART

The literature specific to art in public spaces is recent[10] as well as quite small, and the subject is seldom covered in journals;[11] there is little direct pre-history for the practice (with the exception of murals funded through the Works Progress Administration in the USA during the 1930s, or the work of Mexican muralists such as Diego Rivera) unless its definition is expanded to include the architectural ornament of, say, Art Nouveau,[12] or the Baroque. The context of pre-modern cases, such as the fresco cycles of pilgrimage churches, is too different from that of industrialised society to enable comparison between these cycles and public art.

Before reviewing the literature, two works which illustrate the state of debate in the UK when they were published, one at the beginning of the development, one when it had begun to be established, will be considered: John Willett's *Art in a City*, the published form of his report on art for the city of Liverpool; and *Art Within Reach* (Townsend, 1984), a set of critical and informational essays. *Art Within Reach* will be briefly compared with a collection of essays published at the same time in the USA: *Insights/On Sites* (Harris, 1984).

Art in a City

Art in a City was published in 1967, the year in which, arguably, 'public art' began.[13] It asks how art can be purchased or commissioned for the benefit of a city's population; Willett does not exclude from this the publicness of collections in city museums, nor the heritage of nineteenth-century statuary and architectural decoration. His solution to the non-subject of 'art and society' is to look first at the community in Liverpool. Although it would be argued today that one city is home to many communities differentiated by race, class and gender, Willett's study was pioneering in 1967 in its focus on the reception of art, and it remained isolated in the literature until Sara Selwood's study for the Policy Studies Institute in 1995. Willett describes Liverpool's social condition: 'One sees it immediately in the children . . . off like small rodents on their own private or collective affairs, remote from grown-up control' or 'the visitor continually sees, as he (sic) does not in London or in comparable foreign ports, the symptoms of a population living close to the margin' (Willett, 1967: 4–8). Comparisons are made with Rotterdam, Antwerp and Bremen, highlighting the lingering dereliction in Liverpool but also the city's tradition in the arts, its statuary, diversity of cultural institutions and a few 'striking pieces' of modern works in public settings, such as Epstein's *Spirit of Liverpool Resurgent*, and – 'the best modern Liverpool mural' – a metal relief by Arthur Ballard in a television shop (Willett, 1967: Fig. 66); in a later chapter he refers to the thriving 1960s art scene, which produced work such as Adrian Henri's parody of Ensor, *The Entry of Christ into Liverpool*.

From his observations, Willett derives three inter-related issues: the question of taste, the education of the eye, and the relation of art to society and popular visual culture. One of the best-selling prints in the city's picture shops at the time was Tretchikoff's *Balinese Girl* (probably not a work found in the homes of art critics or arts managers, except as a statement of irony); Willett suggests that whilst such mass-produced prints cannot reflect the locality, they do, on the evidence of works in amateur exhibitions, determine local taste, although he found no clear agreements of taste in gallery art amongst those questioned through surveys.[14] He correlates a lack of formal education with a high rate of rejection of art in general, concluding: 'It is really a very difficult dilemma, for even the most intelligent (and often the most socially-minded) of modern artists tend to skate over these problems – which are not normally talked about except by hostile critics' (Willett, 1967: 137). As a way out of the dilemma, he looks to the work of Hoggart, Hall, Barthes and Eco, who approach art from a concern for its reception, at the same time relating its production to social change and new technologies (Willett, 1967: 235); this framework, Willett argues, softens the division of high and low cultures by looking to media such as film and jazz.

Willett, whilst cognisant of the histories of art in the service of revolution, economic growth and moral rectitude, and guarded on the role of the Arts Council – 'trying to find takers for the kind of art that they themselves enjoy' (Willett, 1967: 239) – accepts that art is 'good' for society, seeing television and the 'Sunday colour magazines', and art as a subject in all levels of formal education, as ways to widen access to it, and he cites various cases of public patronage, such as Zadkine's *The Destroyed City* in Rotterdam and the Middleheim sculpture park in Antwerp, as ways to locate art physically where it might have a wider audience. What he does not go into, and what was not on the agenda in 1967, is the difference between the public realm and public space; nor does he resolve the divergence of popular taste and received taste except by urging risk-taking with works of 'quality', which leads back to the question of by whom 'quality' is determined. At the outset, then, the debate on public art began to open questions which it could not at the time answer; during the past thirty years a vast literature of critical theory, much in translation from European languages, has become available, enabling its readers to extend Willett's problem of the imposition of a received aesthetic into problems of cultural capital, hegemony and representation. In 1967 these perspectives were seldom applied in art history,[15] and it is not surprising that they were not part of the discourse of public art, a term hardly in use in 1967; what is significant, affirming the collusion of public art in the construction of a dominant culture, is that the frameworks which have since radicalised art history have remained largely excluded from discussion within public art. By 1984, the date of *Art Within Reach*, post-structuralism and feminism had already impacted on the 'new' art history and the discourse of mainstream art,[16] but the literature of public art did not move in this direction.

Art Within Reach

Art Within Reach (Townsend, 1984) covers two broad areas: a critical approach to public art; and mechanisms to expand the practice. Willett's essay 'Back to the Dream City' opens the book with some candid statements: 'All that remains lacking is some essential information and thought' and 'Too much of the writing on the subject is uncritical, precommitted to the view that this is a good cause' (Willett, 1984: 7) though he still holds to excellence: 'it was paternalism and elitism that founded our public galleries . . . these positions have to be defended in an age where they are treated as dirty words' (Willett, 1984: 10).[17] His essay raises two kinds of issue which are taken up by other contributors: responsible practice in commissioning, addressed by Peter Dormer through hospitals, shopping centres and schools, by Deanna Petherbridge through public transport and property development, and by Henry Lydiate through law and contract; and the construction of meanings, addressed by Charles Jencks in terms of collaborations, and by John McEwen on community arts. Whilst Dormer, like Willett and Petherbridge, assumes a received notion of quality,[18] McEwen sees the question as more problematic: 'Artistic fragmentation has led to an increasing isolation of artist and audience' which produces a loneliness effectively confronted only by the community artist acting as a catalyst to the creativity of others, but perhaps not assuming the same primacy of aesthetic quality. The community provides the context and hence value of the work, so that the lack of a commonly accessed heritage ceases to inhibit. This glosses over a number of difficulties: the assumptions and visual language which community artists bring with them; that not all community artists have long-term roots in a community, some adopting the mobility of agit-prop and some being opportunist in response to the arts funding system; and that the notion of community is not always transparent to those thought to comprise it. McEwen seeks to link community and mainstream art, seeing community arts as an art-form in itself, 'not just some "artistic" branch of the social services papering over urban decay' (McEwen, 1984: 75); but his anxiety reflects a chasm between the mainstream artworld, including artists who make works for public sites, and community-based artists outside the networks of teaching and exhibiting through which ideas are debated and patronage dispensed.

Jencks, in contrast, states: 'There is no reason in an agnostic society to hire an artist, for what is there to express?' (Jencks, 1984: 17).[19] He rejects sentimental reasons for a 're-marriage' of art and architecture in our secular age, yet takes as cases of past collaboration the 'usual suspects' – Chartres, the Parthenon and St Peter's in Rome – charting a history of the ratio of the costs of art and architecture from 95 per cent for art in ancient Egypt to 50 per cent at Chartres and down to 5 per cent at the Bauhaus, with nothing in Seifert's National Westminster Bank building, which 'isn't even architecture. It's building' (Townsend, 1984: 17). The figures are, he admits, juggled, but the main problems with his argument are the false continuity in the terms 'art' and 'architecture', and the way he sees a society without a unifying religion as if it had no other kind of social cohesion, no networks of sociation and the creation of meaning.[20] Jencks, who seems to project the isolation of the studio onto society, states his 'minimum condition' for a partnership between artists and architects today as a new symbolic programme, but gives little indication of where, outside his own *Thematic House*, this might be found. Brighton writes (later and coincidentally): 'A shared symbolic order is only possible in a totalitarian culture' (Brighton, 1993a).

If *Art Within Reach* stands for the state of debate in the UK in 1984, it suggests that questions such as the relation to identified communities (McEwen) and role of monuments (Willett), are still open. But the debate has not been taken into a more interesting, or more critical or theoretical, area, and many of the contributions advocate rather than consider public art. In the same year, *Insights/On Sites* (Harris, 1984), in the USA, set out a case history of collaboration between Cesar Pelli, Scott Burton and Siah Armajani at Battery Park City;[21] it also covered temporary projects organised by the agency Creative Time, and included a critical essay by Kathy Halbreich on 'The Social Dimension'. The role of artists in the design of city spaces is covered by Richard Andrews in his essay on 'Artists and the Visual Definitions of Cities'. The writers are more optimistic than their British counterparts, who seem under siege by comparison; Anita Contini states: 'This is an exciting time for public art – a time when the process of selecting art and artists for public places is as important as bringing the art into those places' (Harris, 1984: 39), and Halbreich, who refers to W. H. Whyte's work, writes: 'only rarely does the conjunction of site, situation, and shelter even hint at comfort, connection, or community' and accepts a diversity in which there is no single 'public' voice or perspective, a situation which may cause friction as well as celebration (Harris, 1984: 49–50). Experimental work and other art forms – music, dance and theatre – are seen as engaging with the space between art and audience: 'Public art should not be restricted to artworks placed in public plazas but should encompass relationships and dialogue between artists and the public' (Harris, 1984: 47). The book promotes public art by demonstrating what it considers good practice, but does this without closing down the contentions, although a radical critique of art at Battery Park City was not forthcoming until two essays by Rosalyn Deutsche were published in 1991.

ADVOCACY

Art Within Reach, for all its limitations, contributed to a sense of public art having a discourse; what appears to have happened next in the UK is a successful drive, supported by the Arts Council and the funding of public art agencies by Regional Arts Associations, to increase the patronage of public art and improve its 'quality' whilst restricting discussion on public art to areas of administration and funding, on the assumption that public art is 'a good thing' and assuming that the urban developments within which the major opportunities to commission were foreseen, are an equally 'good thing'; a by-product of this is that quality becomes associated with the status of the site and size of the budget for the work, rather than its reception. This took place parallel to a decline in the financial and administrative support for community arts as a result of political changes (such as the abolition of the Greater London Council),[22] further tipping the balance towards public art as an aesthetic embellishment rather than social intervention.

Publications in the later 1980s dealt mainly with the documentation and mechanics of public art. *Art for Architecture*, published for the Department of the Environment, has sections on how to commission art, case studies and a brief historical introduction which sets contemporary pleas for 'art and architecture' within a development from the Art Workers' Guild to the Bauhaus, as if this represented a continuous and hence legitimising history for current practice. Social relevance is one of four 'central issues' in the introduction, but is expanded only by a note on public subscriptions and a section on consultation which concludes: 'The information can be used by the artist or the clients in an imaginative and unstructured manner' (Petherbridge, 1987: 96), which might, given the emphasis on highly structured processes illustrated by complex diagrams elsewhere in the book, be interpreted as 'filing it in the waste-basket', affirming the role of the expert in determining taste; as Willett had written twenty years previously:

> The weakness of the policy of the last twenty years is that it rests on assumptions which are very seldom discussed. Admittedly this saves a great deal of argument, and to those who feel that the assumptions are on the whole the right ones it may seem safer to establish them on grounds of authority, snobbery, faith and (strongest of all) custom, than to thrash them out before a possibly unsympathetic public.
>
> (Willett, 1967: 220)

In contrast, *Going Public* (Cruikshank and Korza, 1988), supported by the NEA, is a more comprehensive handbook, including some critical texts as well as giving documentary material, guidelines, model policies and cases from the percent for art programmes of one hundred and thirty-five cities and states in the USA. Again, a greater confidence in public art in the USA seems to have allowed debate, whilst

in the UK most energy went into securing budgets and making public art acceptable to various art, corporate and government establishments.

Art for Architecture was published in a political climate of decreasing public
expenditure and celebration of 'market forces'; the construction of a business case
for the arts became part of an increasingly professionalised culture of 'arts management' which adopted models of success from other industries, a transition enhanced
by the new concept: 'the cultural industries', which included cinema and rock
music as well as opera, theatre and art. John Myerscough's report *The Economic
Importance of the Arts in Britain*[23] established what seemed like that business case,
demonstrated in the culturally-led return of economic confidence in Glasgow,
which became European City of Culture in 1990.

Publications played a key role in advocacy, legitimising the message in print,
amongst them *Percent for Art: a review* (Arts Council, 1991),[24] a handbook to
persuade local authorities to adopt the policy. Another Arts Council booklet, *An
Urban Renaissance* had already set out in 1989 some of the groundwork in sixteen
case studies of arts projects and arts centres in urban areas, and included the
statement: 'Attractive architecture, landscaping and art in public places enhance
the value of developments for years to come' (Arts Council, 1989: 6), which forms
a basis for the case made in the 1991 publication, which mixes undemonstrable·
claims, such as making places more 'interesting' (to whom?) with the vested interests of the art world in terms of increasing investment in art. It also assumes that
visual art leads the process, although in many precedents it is other art forms,
more participatory in some ways, like music, which were the model. *The Furnished
Landscape*, published in relation to an exhibition at the Crafts Council in 1992,
makes a complementary case for crafts, as street furniture and bridges, in urban
and rural landscapes, though in more discursive essays. The editor writes of 'little
sense of a strong contemporary vision for our communal created spaces' (Heath,
1992: 31) and profiles work in, for example, Swansea and Lewisham, where art
is integrated with urban design.

As a result of advocacy, a significant number of local authorities, around 40
per cent by 1993 according to Selwood, adopted a public art policy of some kind;
she reports that not all were implemented, usually due to financial pressures
(Selwood, 1995: 299), but a report from the University of Westminster (Roberts
et al., 1993) suggests that the public sector commissions around three times as
much art as private-sector property development. Two particular outcomes of 1980s
advocacy have affected the parameters of debate on public art: an expansion of
the number of public sector administrative posts in the field; and an increase in
the levels of budget and mix of public and private sector sources of funding for
commissions.

The complacency of the received case for public art was, however, exploded
by Selwood's *The Benefits of Public Art* (Selwood, 1995) – an ironic title – and her

earlier essay for *Art in Public*, in which she contrasts 'accepted wisdoms' with 'the derogatory colloquialisms public art inspires [which] suggest a different story — "turds on the piazza"; "parachute art"; "developers' equine"; "corporate baubles"; "lipstick on the face of a gorilla"' (Selwood, 1992: 11). In *The Benefits of Public Art*, Selwood sought public and professional responses, sometimes working through focus groups, to selected commissions and projects, such as Henry Moore's *Arch* in Hyde Park, art in Broadgate, town artist Francis Gomila's work in the refurbishment of Smethwick High Street, and Centenary Square in Birmingham. The report exposes the continuing dominance of taste in public art management and the expectation that audiences rather than artists should make concessions, the needs of the community and environment being secondary. Her executive summary states: 'None of the case studies had been formally evaluated by those involved. The specific criteria according to which they might have been evaluated were rarely identified as such' (Selwood, 1995). She describes some of the stated objectives as self-fulfilling or purely administrative and her recommendations include a greater openness and accountability in all aspects of commissioning, and an on-going process of evaluation.[25]

Throughout the late 1980s and into the 1990s, a critical debate was taking shape in the USA, in part around issues of user participation — Jerry Allen asked in 1985 how art addresses the heterogeneity of society: 'It enters an environment which removes art from the privileged and specialised status of the museum into a much wider realm' (Allen, 1985, cited in Cruikshank and Korza, 1988: 246) — and in part within the context of social criticism from which Deutsche was writing and activist practices were emerging. This does not mean that in the USA there was no dialogue on administrative questions; Lacy recalls that public art had become a recognisable field on which conferences were organised, but also that all those involved, from artists to administrators and critics, looked at the progression from museum to public places as sites for art in context of a formalist framework (Lacy, 1995: 24–5).[26] Amongst those demonstrating this is Beardsley, whose description of art in Battery Park City is considered in the next chapter.

CRITICISM

A problem in writing critically about public art is the lack of the well-rehearsed models which exist for art in galleries — the article based in art historical method, showing the derivation and development of the work; the interview with a 'star' artist; and the exhibition review, as much, following Baudelaire and Ruskin, about the reviewer's taste or state of mind as the art. Neither has art in public spaces been the subject of the psychoanalytic reading of Peter Fuller's *Art and Psychoanalysis* (Fuller, [1980] 1988). But how, for instance, would a review be

written of a piece of street furniture such as a seat or litter bin?[27] And whilst
factors such as budget and management process are important in public art, to
write a review of an exhibition beginning with the price list and moving onto
the dealer's business arrangements or the museum's sources of funding would be
seen as derogatory, or subversive, as in the work of Hans Haacke.

Brief examination of two texts shows a divergence of approach to the problem:
one an extension of conventional art criticism; the other an example of critical
writing closer to the model of cultural studies.

Bottle of Notes

Richard Cork's 1995 review for *Modern Painters* of *Bottle of Notes*, a public sculpture
in painted steel to commemorate Captain Cook, by Claes Oldenburg and Coosje
van Bruggen, aims to elicit a sympathy for the artists and the civic authority in
Middlesbrough who commissioned the work; he begins: 'Making monumental
sculpture for a civic site is an enterprise fraught with potential pitfalls' (Cork,
1995: 76). Cork sees public art as deriving its value from individual creativity
and the freedom of the artist in the studio, a standard modernist view, but: 'The
outcome could become so divorced from its maker's authentic imaginative concerns
that banality sets in' (ibid.: 76); he takes an art historical approach to derivation
and context, in the bulk of the text going through early sketches for the work,
referencing a story about a manuscript in a bottle by Poe, which seems as much
a source as Captain Cook's notebooks (from which the squiggle forms in the work
are taken), and situating *Bottle of Notes* in terms of Boccioni's *Development of a
Bottle in Space*, 1912, and Tatlin's tower – 'The whirling lines . . . have a whiplash
energy reminiscent of Tatlin's Monument to the Third International, a soaring
model which encapsulates the utopian dynamism of the Russian avant-garde' (ibid.:
80). But, does this mean the good folk of Middlesbrough are communists? Or is
the ideological programme behind Tatlin's constructivism ignored in favour of the
formal qualities of a modernist critique? The question of Cook's role in 'discov-
ering' other people's islands is not entertained, although the work could be seen
as heritage culture; and whilst there are references to the work's local fabrication,
its siting near new municipal offices and possible responses within a changing
urban landscape, the style of Cork's review is somewhere between art history – a
set of precedents through which to legitimise the work as art – and reassurance
of the patrons in their desire to put themselves on the international 'culture map'.
He ends:

> Even as the people of Middlesbrough sort out their responses to Bottle
> of Notes, it serves a far wider purpose, too. For the town can be proud
> of the fact that such a distinguished work will quickly acquire an

international reputation, and that nobody else in Britain has been adventurous enough to commission a sculpture from Oldenburg and van Bruggen before.

(Cork, 1995: 81)

No doubt the 'people of Middlesbrough' are still at it, 'sorting out their responses' in an orderly way, irony here, hegemony there, waste-of-money in the cupboard, hermeneutics in the public bar; Cork assumes a simplicity in the relation of art to a site which is neither gendered nor ideologically constructed; but that is his role as a compliant modernist critic.

Out of Order

Patricia Phillips' article 'Out of Order: – the public art machine', for *Artforum* (Phillips, 1988), contrasts the claims for public art as a specialist discipline with its lack of theory, and is, perhaps, the beginning of a new critical approach to public art which is still developing. Her main point is that public art limits its notion of what constitutes being public to public space:

> One basic assumption that has underwritten many of the contemporary manifestations of public art is the notion that this art derives its 'public-ness' from where it is located . . . The idea of the public is a difficult, mutable, and perhaps somewhat atrophied one, but the fact remains that the public dimension is a psychological . . . construct.
>
> (Phillips, 1988: 93)

Her solution is an engagement with public issues, which intersect with personal interests and collective values, from which public art might be freed from the linked agendas of developers and city officials who have commissioned art in public spaces but not created access to the public realm: 'Thus public space has served as a great new incentive – not to be "public", however, but to satisfy far more profit-motivated market objectives' (ibid.: 93). She supports temporary projects on the grounds of the greater risks which can be taken, and questions community representation on selection panels, her juxtaposition of community and 'general public' being intentionally problematic:

> the ideas of the local community and of the general public are put into an adversarial relationship . . . This peculiar endorsement of community opinion, sometimes at the expense of larger public concerns, subtly yet effectively affirms a notion of what I would call 'psychological owner-ship', at the same time that it refuses to ground that notion in any terms other than geographic.
>
> (Phillips, 1988: 94–5)

In a later article on Peggy Diggs, Phillips draws attention to a need, particularly for women artists and reflecting the position that 'the personal is political', to cross the boundaries of gendered public and private domains. Towards the end of the *Artforum* essay, she states that to go beyond an understanding of the new urban landscape 'as a geographic grid' means that the aspiration of public art expands 'to include the "invisible" operations of huge systems and the intimate stories of individual lives' (Phillips, 1988: 96). Whilst Cork was concerned to legitimise public art as modern art, Phillips is concerned to problematise it.

Critical sources and issues

Since the late 1980s, an increasing volume of critical writing on public art has appeared, including Arlene Raven's *Art in the Public Interest*, first published in 1989 (Raven, [1989] 1993). Raven writes:

> Activist and communitarian, art in the public interest extends the modes of expression of public art of the past several decades. The new public-spirited art can . . . critique . . . the uneasy relationship among public artworks, the public domain, and the public.
>
> (Raven, [1989] 1993: 1)

She emphasises process above object, and introduces the strategy of group working which includes non-artists – a list from clowns to shamans and laundry experts – related to specific localities and their publics. Amongst the contributors, Lacy compares the reception of art in public spaces to that of mass media 'spectacles'[28] which manufacture a desire for constant innovation and stimulation regardless of ethical decisions, so that 'novelty . . . is taken as proof of authenticity' (Lacy, 1993: 298). Her own work, such as *The Crystal Quilt*, a performance piece involving 430 elderly citizens of Minneapolis in 1987, sought to set up a cultural model of empowerment.

Key sources for the emerging debate, in which 'conventional' public art became increasingly characterised as collaboration with dominant state and corporate, national and global structures of power, were two multi-author volumes produced by the Bay Press in 1991: *If You Lived Here* edited by Brian Wallis, and *Out of Site*, edited by Diane Ghirardo,[29] both including essays by Deutsche. Three further volumes (also multi-authored) appeared from the Bay Press in 1995: Lacy's *Mapping the Terrain*, Felshin's *But is it Art?*, together with a volume dealing specifically with the *Culture in Action* project curated in Chicago in 1993 by Mary Jane Jacob. A reappraisal of modernism's socially disastrous consequences (mapped onto wider environmental concerns) was undertaken by Suzi Gablik in *The Reenchantment of Art*,[30] in which she sees the success of art as determined by its ability to usher in a paradigm-shift, out of Cartesian dualism into an art of participation, social relatedness and ecological healing. Contributions around issues of identity, from both practitioners and academics, in *The Subversive Imagination*, edited by Carol Becker, include a retrospect on her practice and its ideological context by Martha Rosler, and Becker's own essay on Marcuse's late text *The Aesthetic Dimension*. Dolores Hayden's *The Power of Place* describes projects to give form to the memories of place of people in minority communities in Los Angeles. The other key work, the only one published in the UK and concerned with wider futures than simply those of public art, is *Mapping the Futures*, which includes contributions from geographers Doreen Massey and David Harvey, sociologists Mike Featherstone and Ruth Levitas, and the philosopher Gillian Rose. Jon Bird contributes a chapter on the London Docklands development as 'Dystopia on Thames', which is followed by an account by artists Peter Dunn and Lorraine Leeson of their work with the communities disaffected by the development. Bird cites Laclau, whilst Dunn and Leeson cite Gramsci, Habermas and Said, and the book as a whole situates certain kinds of public art, and certain kinds of development, in terms of critical theories of society. A recent contribution is *House*, a collection of essays on Rachel Whiteread's work of that name, including texts by Bird and Massey.

Taking this range of books, two characteristics emerge: that all except one are multi-authored; and that all those include contributors from more than one academic discipline and/or from theory and practice.

Histories

Mapping the Terrain includes brief overviews of the work of a range of new genre public artists and offers two histories of public art:

> One version of history, then, begins with the demise of what Judith Baca calls the 'canon in the park' idea of public art – the display of sculptures glorifying a version of national history that excludes large segments of the population. . . . In the most cynical view, the impetus was to expand the market for sculpture
>
> (Lacy, 1995: 21)

and:

> An alternative history of today's public art could be read through the development of various vanguard groups, such as feminist, ethnic, Marxist, and media artists and other activists. They have a common interest in leftist politics, social activism, redefined audiences, relevance for communities (particularly marginalized ones), and collaborative methodology.
>
> (ibid.: 25)

Lacy, like Gablik in *The Reenchantment of Art*, constructs through the artists and projects profiled a milieu and a moment, in which the mainstream notions of art are irrelevant – Alan Kaprow, working in schools and communities during the 1960s, writes of *Project Other Ways*, which involved students at a California public school, that it was: 'intent on merging the arts with things not considered art' (Lacy, 1995: 154). Several other essays relate to, and artists profiled come from, marginalised, non-white communities – Native American Hachivi Edgar Heap of Birds, for instance – whose histories have been stolen.

The sense of a milieu with roots in the alternative practices of the 1960s is also given by *But is it Art?* – twelve critical essays covering activist art over three decades and establishing this history as a counter to the 'canon in the park' to 'modernist sculpture in the park' model. Phillips contributes essays on the 'maintenance activity' of Mierle Ukeles (whose work had previously been brought to notice by Gablik), and on Peggy Diggs; Felshin writes that twentieth-century art was produced, distributed, and consumed in context of an art world guided by personal expression, but that from the late 1960s, changes in art reflected changes in the 'real world' (Felshin, 1995: 10). But it is not a history without difficulties, and Felshin sees the crises of identity, even demise, of some groups as part of a continuing flux. Rosler in 'Place, Position, Power, Politics', in *The Subversive Imagination* states: 'patterns of behaviour and estimations of worth in the art world are more and more similar to those in the entertainment industry . . . [which] promises that everything will go down easy, even mass murder' (Rosler, 1994: 57), whilst art tends to lecture its publics. Hal Foster had earlier drawn attention to the co-option of site-specific art to the mainstream[31] and Rosler's photomontages are now sought by dealers. But her memory of the 1960s probably stands for many 'new genre' artists: 'When I finally understood what it meant to say that the war in Vietnam was not "an accident", I virtually stopped painting and started doing agitational works' (Rosler, 1994: 58).

AN END OF CONTRADICTIONS?

New genre public art in the 1990s resolves the contradiction of public art by re-determining 'public' as the 'space' (or 'time') of public issues while subverting the gendering and separation of the public and domestic realms. Phillips sees institutional public art, in contrast, as excluded from real interaction: 'In spite of an appealing, ambitious agenda to make art available, if not central, to the lives of individuals and communities, public art remains theoretically and practically marginalized' (Phillips, 1995: 60).

The critical project, then, of Phillips, Lacy, Gablik and others is to situate alternative practices in a wider urban and social, for Gablik ecological, discourse. There remain two areas of difficulty: first, the effectiveness of public art in building strategies for intervention but also practices of participation which are not received as artworld 'lectures', being a catalyst for empowerment in place of liberal reform, and resisting co-option into contemporary art's resemblance of the entertainment industry; second, the extent to which a redefinition of public art implies a total re-visioning of western structures of value and the adoption of a position within the ecological discourse – 'ecosophy' – as proposed by Gablik.

Gablik is perhaps the most radical in her position, foreseeing a paradigm shift, an overturning of the 'perception of the world that we have been taught', in which Cartesian dualism is discarded and art 'will begin to redefine itself in terms of social relatedness and ecological healing' (Gablik, 1991: 27). Is it the end of the Enlightenment project of·Progress? Can there be a reclamation of the practices of everyday life in a new kind of sociation based in connectedness?

5

ART IN URBAN DEVELOPMENT

•

INTRODUCTION

Urban development, through both public and private sector initiatives, is one of three principal markets (with public transport and health services) in which the commissioning of public art has expanded. But urban development takes many forms, not all socially beneficial or synonymous with the regeneration of local economies.

Advocacy for public art has sought opportunities to integrate art in major capital schemes, often through the employment of artists in the design stage of buildings, but it conflates all types of development regardless of their social or environmental impact, whilst there is no systematic evidence that public art has beneficial effects for urban communities, who may be marginalised rather than regenerated by development.

Just as development takes contrasting forms, so the arts in this context vary in scale and intention; some projects are rooted in their locality and its communities, or address the decline of a defined section of the built environment, as, for example, Francis Gomila's work with local shop-keepers in the regeneration of Smethwick High Street;[1] other projects claim a location on an international 'culture map', as with *Bottle of Notes* in Middlesbrough, or are in effect collections of 'blue-chip' modern art, as in the Broadgate development in London. Arts policy, however, in forms such as Percent for Art, makes no distinctions between kinds of development, accepting the contradictions between the responsibility of the public sector in a liberal society and the non-accountability of some developers to interests other than those of global finance.

This chapter examines the ethical aspects rather than mechanics of public art policies.[2] It charts the derivation of a case for the support of public art from that made for the economic importance of the cultural industries, and scrutinises the strategy devised for art in the Cardiff Bay development. It continues in examining the model of criticism formulated by Rosalyn Deutsche in relation to art in Battery Park City, contrasting art in corporate development with the community-related art project in Sunderland's re-development zone. It is a long chapter, in some ways the fulcrum of the book's general argument that the absence of a critical or ethical framework for public art leads to its complicity in urban dis-ease.

A CASE FOR ART?

There are three 'shells' of advocacy inside which this problem is encased: a business case for public art in development schemes: this derives much of its content from advocacy for the arts in general as contributors to economic life; that is in turn set within the case for urban development as a means to a 'better world'; inside the 'can' of the advocacy are at each level many 'worms' in the form of social failures.

Social zoning

Images of the metropolis are polarised: new developments are surrounded by areas of residual decay to the inhabitants of which the 'benefits' of development do not extend. These hitherto stable if impoverished and 'disorderly' communities are re-classified by development as marginal. Corporate development and affluence are put into focus by 'acceptable levels' of deprivation in housing, health care, education, transport and culture. In his 1988 address to the Royal Town Planning Institute, Francis Tibbalds spoke of 'an environment of private affluence and public squalor . . . with no effective means of controlling it' (cited in Selwood, 1991); Richard Rogers writes of the London Docklands development: 'the concept of balanced development was all but ignored in favour of policies geared to facilitate fast, financially attractive exploitation of the area' (Fisher and Rogers, 1992: xxvi), and asserts in a later text that 'The dominant philosophy of the next century will be sustainability . . . This relates to a public responsibility rather than the predominant factor of today, which is private greed' (Rogers, 1995: 3).

Development has aspects which are more complex than the juxtaposition of greed and responsibility. It produces sharply delineated (geographic and conceptual) zones of 'success' which define counter spaces of 'failure', and separates monolithic corporate culture, which increasingly stands for 'the city', from the diversity of street cultures, leading to a process of abjection in which, for example, the homeless are identified with pollution and their 'plight' is made to seem either voluntary, or a misfortune which cannot be helped, or a crime against society. David Sibley writes of:

> visions of purity and pollution where the polluting are more likely to be
> social, and often spatially, marginal minorities [whilst] media represen-
> tations . . . draw on the same stereotyped images of people and places

> which surface in social conflicts involving mainstream communities and 'deviant' minorities.
>
> (Sibley, 1995: 60)[3]

Rosalyn Deutsche begins her critique of art in Battery Park City by noting the neutralisation of eviction (the term 'homeless' already shifts meaning from a purposive act to a misfortune) in the city's dominant culture; and Saskia Sassen, writing on corporate cultures and their borders, argues: 'The slippage is evident: the dominant culture can encompass only part of the city. And while corporate power inscribes these cultures and identities with "otherness" thereby devaluing them, they are present everywhere' (Sassen, 1996: 188). The dream city, it seems, is for the saved. The others wait outside.

In the UK, the widening of social divisions and increasing political influence of development takes place alongside a trend to heritage culture; Jon Bird writes of: 'a continuing nostalgia for a mythical past of tranquillity and order running deep within the political and social fabric of Britain' and cites an extract from the publicity material of the London Docklands Development Corporation (LDDC) which speaks of 'London's bustling docks and busy river' as 'jewels in Britain's commercial crown', with their 'exotic cargoes and evocative aromas from far-flung corners' (Bird *et al.*, 1993: 120–1). Forms of this nostalgia include: the appeal to past 'greatness' as a device for political unity; the preservation of residual forms of the past in industrial and 'everyday life' museums;[4] and the adoption of post-modern vernacular styles to create an illusion that some parts of a development 'belong' to a sense of place which the development has destroyed. Perhaps *Cincinnati Gateway* is a version of heritage culture, too, its 'map' of the past bland because the artist's research into local history excluded contended definitions of locality; personal memories of place, on the other hand, may resist homogenisation. But it is the artificial sense of locality of, for instance, Tudor-veneer supermarket buildings, the deception thin enough to be cynical, in which heritage is most conflated with corporate image.

Urban development, either futuristic towers of glass or homely vernacular, sets up an adversarial social model: 'good' affluence against 'defiling' deprivation, on a colonial model of 'developed' and 'developing' countries, the perception becoming a justification for the value-laden process which produces it, just as Burgess' concentric ring model was re-coded from description to prescription. Art in development aids this socially divisive process by aestheticising it – and a little controversy over art is sometimes not unwelcome if it distracts attention from social issues; but in ignoring the social impact of development, art is complicit in the consequent social fragmentation.

Gentrification

A model of arts-led development detrimental to existing communities and which follows a longer history of the purification of urban space (discussed in Chapter 1) is the gentrification of areas through arts uses, leading to an increase in property values which drives out the residual community. Artist Martha Rosler writes of New York's SoHo that artists were a pivotal group in easing the return of the middle class to the area, although artists themselves were displaced by the wealthy clients who followed them into the newly chic neighbourhood (Rosler, 1991: 31);[5] and Savitch writes on London's Covent Garden that a new social structure has been interspersed with the old, as shops are replaced by wine bars, so that 'community preservation may not just be preservation for the community' (Savitch, 1988: 222–3). Gentrification 'clarifies' an area, substituting a simplified and constructed identity (or representation) for the muddle of ordinary use. This is a contemporary form of the purification of parts of the city for bourgeois life which took place in the eighteenth and nineteenth centuries.[6] It follows that gentrification usually concerns neighbourhoods which can be easily defined by some form of appropriation of characteristics, like SoHo's iron architecture and Covent Garden's open spaces; new development, on the other hand, begins from the sort of site clearance of which Le Corbusier dreamed.

But development, despite subsidies from taxation intended for public benefit, may fail to regenerate a locality; industry may create jobs which are highly specialist and not filled locally, whilst infrastructures lag behind the process and pre-existing, if less legible, networks of support are obliterated; or development may take the form of a corporate enclave isolated from its geographical surroundings – geographical proximity falters in context of global networks of electronic communication. Peter Ambrose argues that development is purposeful, subject to planning and cost-benefit analysis, but that whilst the mechanisms may work in their own terms, the key political question is who controls the process: 'the initiators and organisers of change, the active modifiers of the built environment, are not usually conceived of as "us" at all but as "they" . . . usually some shadowy undifferentiated "they"' (Ambrose, 1994: 6–7). Development, then, and particularly following the spectacular collapse of Canary Wharf,[7] has an image problem. Art and architecture address this by providing 'beauty' (conventionally associated with truth and goodness), whilst the commissioning of art through intermediaries – 'art experts' – replicates that sense of a 'they' who remotely determine the form of the city.

The image of development in the publicity material of schemes such as Docklands and Battery Park City is achieved through the creation of static vistas, not the regeneration of communities or sustainable webs of local businesses. As in most architectural photography, people are absent from the images used, as in the considerations of developers, or represented by stereotypes. Some of the justification advanced for this is in the more widely accepted contribution of the cultural industries to local economies and supposed, if remote, 'trickle-down' effects of development.

Cultural industries

The importance of the cultural industries, particularly the performing arts, in economic viability or regeneration is set out in John Myerscough's 1988 report *The Economic Importance of the Arts in Britain*, which established through studies of Ipswich, Glasgow and Merseyside the attraction of a cultural profile for companies looking at sites for relocation. Peter Rodgers summarises: 'It was hard to find a leading businessman in Glasgow who did not have an involvement in the arts' (Rodgers, 1989: 61) and 'The arts were one of the slender threads holding together Merseyside as a viable region, according to many of the businessmen interviewed' (Rodgers, 1989: 62) whilst 'companies [in Glasgow] took thriving culture as evidence of a dynamic self-confident community' (Rodgers, 1989: 64). A survey of middle managers produced a 69 per cent average (from the three cities) rating for museums, theatres, concerts and other cultural facilities as 'important' in their environment; access to the countryside scored 93 per cent, and spectator sports 22 per cent. This establishes, though related to one strata in society, a case for a cultural profile as a way to attract and retain corporate investment, though models cannot easily be mapped from one city to another and a cultural profile (which depends on there being publics for culture) cannot be manufactured overnight or bought-in as a package. Nevertheless, some cities at which Myerscough did not look lend supporting evidence – the Temple Bar (cultural) district of Dublin and (mixed use) Castlefield in Manchester, amongst others.[8] The Newcastle Arts Centre, on a smaller scale, creates a cultural economy in one street,[9] involving both architectural heritage, training in restoration, and contemporary crafts and performance; there are several examples of redundant industrial buildings renovated as artists' studios, like the Custard Factory in Birmingham.[10] What this body of evidence concerning *processes* of regeneration through the cultural industries does not do is establish a case for capital *acquisitions* of contemporary visual art in isolation from a cultural matrix involving a diversity of art forms, venues and publics; the arts profile to which Myerscough refers tends to be the result of a critical mass in which the performing arts are the largest element – music in Liverpool, drama in Glasgow's Mayfest. This inconvenience seems not to inhibit the advocates of public art, whose conceptual maps appear remarkably neat and tidy.

Arts policy as advocacy

In this context, Luke Rittner, as Secretary-General of the Arts Council, wrote in the Foreword to *An Urban Renaissance*: 'Urban renewal continues to be high on the national agenda. The architecture and quality of life in our cities are subjects of debate throughout the country . . . The arts are making a substantial contribution to the revitalisation of our cities' (Arts Council, 1989). This publication profiled sixteen projects across the country which it saw as contributing to the revival of business confidence, the quality of environment, and community participation in cultural activities. The text states that 'Arts activities provide a community with a focus and increase its sense of identity . . . an increased awareness of the community's needs, a determination to achieve change'; it mentions festivals and cultural tourism, and asserts that 'The arts often serve as the main catalyst for redevelopment', noting (unspecified) cities in the USA where cultural districts have been in the forefront of regeneration. The projects illustrated can be categorised as four multi-arts venues, six performance arts projects, two media projects, and four visual arts projects, two of which (the extension of the Ferens Art Gallery in Hull and the proposed European Visual Arts Centre in Ipswich) are buildings, the others being the integration of art in the redevelopment of the Swansea Maritime Quarter and the community arts group Nottingham City Artists. Of these, only Swansea is directly relevant to public art, though the text includes an appeal to developers on the grounds that, alongside architecture and landscaping: 'art in public places enhances the value of developments for years to come' (Arts Council, 1989: 6).

What seems to have happened is that an enthusiastic visual arts lobby appropriated the case for the arts in regeneration and applied it to public art in development, despite that (modernist) visual art lacks the capacity for wide participation provided by, say, street theatre, and is often opaque in meaning to non-art publics; Mike Featherstone, seeing the arts through the discipline of sociology, writes:

> artistic and intellectual goods are enclaved commodities whose capacity to move around in the social space is limited by their ascribed sacred qualities. In this sense the specialists in symbolic production will seek to increase the autonomy of the cultural sphere and to restrict supply and access to such goods, in effect creating and preserving an enclosure of high culture.
>
> (Featherstone, 1995: 23)[11]

Art and architecture, then, have coincidental interests which extend from their shared professional ideologies and align them to the needs of developers to create enclaved 'places', but set them against a model of the city in which its dwellers determine or at least influence its values and forms. Whilst at a local level the cultural industries may assist regeneration, though with the danger of gentrification, the corporate development in which public art is more often commissioned, as at Battery Park City, London Docklands or Broadgate, represents global interests. Sassen writes: 'Today [cities] are transnational spaces for business and finance where firms and governments from many different countries can transact with each other, increasingly bypassing the firms of the "host" country' (Sassen, 1996: 189). Those firms and governments, as financial institutions, developers and Urban Development Corporations, currently re-determine the city amidst its ruin, bypassing the residual (and immigrant) populations. One of the means they acquire to do this, if in a secondary role, is international modernist art.

Social and ethical considerations are not on the agenda of most of the public art agencies established by the Arts Council through its regional structure, as a management infrastructure for the business of public art. Following a report in 1990 from a steering group on the Percent for Art policy (Arts Council, 1990) which met for two years, chaired by architect Richard Burton, the Arts Council commissioned *Percent for Art: a review* (Arts Council, 1991) as a handbook to persuade local authorities to adopt the policy, setting out a case for commissioning public art which has become standard:

> To make a place more interesting and attractive.
> To make contemporary arts and crafts more accessible to the public.
> To highlight the identity of different parts of a building or community.
> To increase a city's/county's/or company's investment in the arts.
> To improve the conditions for economic regeneration by creating a richer visual environment.
> To create employment for artists, craftspeople, fabricators, suppliers and manufacturers of materials, and transporters.
> To encourage closer links between artists and craftspeople and the professions that shape our environment: architecture, landscaping, engineering and design.
>
> (Arts Council, 1991: 16)

Advocacy for public art provided a new market for art and arts management, not as an alternative in response to the recession which occurred slightly later in the art market, but in association with the property boom; Zukin writes of the 1980s: 'the boom in these sectors of business services [finance, insurance, real estate] . . . influenced sharp rises in the real estate and art markets in which their leading members were so active. Investment in art, for prestige or speculation, represented a collective means of social mobility' (Zukin, 1996: 45); this is in keeping with Selwood's statement that 'The fact that works by such prestigious artists as Richard Serra, Jim Dine, and George Segal are integrated into Rosehaugh Stanhope's Broadgate development doubtless contributed to attracting major American, European and Japanese companies to locate there' (Selwood, 1992: 21).

Property developers and urban planning authorities in the UK were subject to a campaign by the Arts Council and public art agencies, in which the Percent for Art mechanism was a central strategy, along with a plea for artists to be involved in the design stage of buildings. The campaign was more successful in persuading local authorities (including Development Corporations) and government departments relocating out of London[12] than private-sector property developers to adopt public art policies;[13] at no time did it appear likely that a national Percent for Art policy would be adopted, or that the concept of 'planning gain' would be used to impose art on developers.[14] Other aspects of the support of visual art in urban regeneration included the inclusion of temporary sculpture parks in the garden festivals at Liverpool, Stoke, Glasgow, Gateshead and Ebbw Vale, between 1984 and 1992; and increasing support for the infrastructure of public art managers to establish their work as a specialism.

The difficulties with the Arts Council's policy are, apart from the mix of undemonstrable claims and artworld perceptions: first that it applies to the whole spectrum from community regeneration to mega-corporate development, without differentiation between corporate greed and the public good; second that it privileges visual art and the role of intermediaries, not so much the guardians of art 'trying to find takers for the kind of art that they themselves enjoy' (Willett, 1967: 239) as making jobs for the kind of people they themselves are; and third that Percent for Art policies favour a narrow interpretation as the commissioning of permanent fine or applied art for new buildings. Massey has argued that the privileging of the visual sense 'allows most mastery, in part deriving from the very detachment which it allows and requires' (Massey, 1994: 224), and this may be part of its appeal to developers, along with the investment value of permanent works. There are two other difficulties: the lack of thinking through the question of encouraging the use of public space, and the lack of evidence on lending a competitive edge to developments. *Percent for Art: a review* states: 'Commissioning bodies argue that good art encourages greater use of public places and increases individuals' sense of security' (Arts Council, 1991: 17); presumably security is a function of numbers (Jacobs, 1961), but, there are questions as to who constitutes the publics to be encouraged, and it could be argued that provision of seating is more important than art – as observed by W. H. Whyte, whose examples of the most popular public space include Greenacre Park and Paley Park in New York, both designed with water features by competent landscape architects but having no public art (Whyte, 1980). The Arts Council case continues 'art on public view gives an impression of social, cultural and economic confidence . . . may also increase land values and the sale or letting price of a development' (Arts Council, 1991: 18); the University of Westminster report of 1993, *Public Art in Private Places*, does not support this to a useful extent – positive responses to buildings highlighted modernity and the provision of semi-public space such as an atrium.[15] Whilst 81 per cent of respondents in companies which had relocated agreed that art enhanced the status of the development, this factor was fifth out of six factors influencing the choice of a building;[16] the report summarises: 'there was a weak link between an occupier's decision to take a tenancy in an office development and the presence of public art', although almost two thirds of respondents linked art to the attractiveness of the building (Roberts *et al.*, 1993: 24). Generally, the 'added value' of public art in property development appears weak, yet major schemes such as London Docklands, Broadgate and Cardiff Bay have commissioned art, suggesting either unusually successful advocacy, or a particular motive in such cases which does not apply to individual buildings within an established urban area which already has cultural attractions, perhaps because the abolition of the past involved in new developments requires a visible inscription of the future, or at least of something distinctive.

Development and regeneration

There are differences between 'urban development' and 'urban regeneration' which the case for Percent for Art ignores. Development means building presented as in the public good ('creating jobs'); it is lent state support in the form of subsidies, tax breaks, visual art expertise from subsidised sources, and a reduction in the effectiveness of planning mechanisms designed to deliver, in other circumstances, democratic accountability ('cutting through red tape');[17] and it sometimes involves 'signature' architecture, or at least an impressive and futuristic skyline seen from across the water.[18] Regeneration means creating sustainable economies, and it includes the means of sociation – a sustainable sense of neighbourhood and 'street life'. One of the roles assigned to visual art is to give the impression of the difference, whilst in actuality having a negligible input to the local economy and lacking the capacity of the performing arts to engender sociation. This imposes on public art the burden of accepting the implicit values of development, which are in general profit and the extension of the power of corporate interests; and the weight of contradictions involved in this.

A particular kind of contradiction, between development and ecology, is found in the policy for art in the redevelopment of the dock area of Cardiff. Like the London Docklands, this scheme is geographically adjacent but separated in ambience and land use from the city, and like it again it abolishes the past of the site, substituting 'bay' for 'docks'; the proposed opera house (a symbol located between the elitism of the most expensive art form and the heritage notion of a 'land of song') was denied lottery funds in favour of a sports stadium. Meanwhile, several sculptures, such as Eilis O'Connell's *Secret Station* and Pierre Vivant's roundabout piece of street signs re-coded as art, have been sited in the redevelopment zone. In terms of attracting funding for public art it has been successful, and Cardiff Bay Arts Trust, the management mechanism established to handle art commissions in the development, has won an award.[19] The series of commissions followed a report by a team of public art managers, which epitomises the territory of contradiction in which a mechanistic approach locates public art policy.

CARDIFF BAY

The *Strategy for Public Art in Cardiff Bay* ('the *Strategy*'), written in 1990, proposes an arts plan costed at £30 million, £16 million to be provided through a percent for art scheme enforced by the Development Corporation; it makes the usual business case for art on the grounds of its perceived role in urban economic regeneration; a possible source of the *Strategy*'s assertions is the claim in *An Urban Renaissance* that the arts are effective in attracting visitors to a town, enhancing the visual quality of the environment and providing a focal point for community pride. The *Strategy* translates arguments from the cultural industries and the arts in general to public art, claiming that the visual arts 'have usually spearheaded the process of urban regeneration' (Cardiff Bay, 1990: 15), for which no evidence is given. A number of art establishment notions are incorporated: 'The most highly regarded Art in Public Places Programmes . . . have aimed uncompromisingly for the highest quality artists and artworks' – here American cities are named: Seattle, Los Angeles (before the riots), Battery Park City (Cardiff, 1990: 16); and 'quality' can be achieved through the appointment of experienced visual arts consultants like its authors; the thinking is reminiscent of Willett's plea for 'risk-taking' in his 1967 report on Liverpool. Most of the report is standard reading within the advocacy of public art, defining site as geographic, beginning from an artworld perspective in which the Bay is a site for art, like a big new gallery.

The most contentious recommendation, which exposes the value-free ethos of the report, is that the 'entire barrage' (seen by ecologists as devastating for the bay's wild life) should be 'a work of art' (Cardiff Bay, 1990: 22) produced through a collaboration of artist, architect and engineer. This seems to contradict the claim that art can 'Improve and enhance all visual aspects of the environment of the Bay for the benefit of those who live, work and play there' (Cardiff Bay, 1990: 19), obviously not including the birds, but also excluding those citizens who regard the sight of wildlife as pleasant and believe in the responsibility of citizenship to safeguard it. The suspicion arises that the main purpose of the barrage is to 'clean' the bay to provide a purified vista for developers, making another of the sharply-demarcated spaces to which Sibley refers in *Geographies of Exclusion*. A recent account describes the 'new' Bay as 'the bare marsh replaced by even more barren brick piazzas, the greedy gulls by developers, tidal slime by cars, nature by man' (Anon, 1995: 10). In contrast, David Reason has argued (in another context): 'We must never again forget that before a plant is of value to me, it is of value to itself, and that the barest desert supports its bounty of life both directly and indirectly. In the syntax of human history the declension goes: dead lands, dead beats, police state' (Public Art Commissions Agency, 1990: 57–8).

The *Strategy* shows professionalism, but also an elision of moral and social issues in favour of optimistic art futures, thus denying a reading of art as liberating through a capacity to imagine alternative futures. Against the dogmatic assertions of the Cardiff Bay strategy we could contrast *Visual Dallas* (Dallas, 1987), which begins with an evaluation of the problems of the city, and the importance for its 'liveability' of good public space, citing W. H. Whyte.[20] Art is then introduced to the agenda of 'creating a city which looks good, feels good, and is accessible', whilst 'the resource activated by public art programs that makes the biggest difference is the public itself . . . they assume their rightful role as members of the team' (Dallas, 1987: 1.7–1.10). This does not preclude the dominance of corporate inscriptions on urban space, and words can conceal unspoken interests, but *Visual Dallas* does at least begin with people; the *Strategy*'s preface states: 'Our aim has been to propose an artistic and financial Strategy inspired wholly in response to the special qualities of Cardiff Bay and the scale of opportunity offered by its redevelopment' (Cardiff Bay, 1990). This is an example of the kind of neutralisation on which Sassen writes:

> The dominant narrative presents the economy as ordered by technical and scientific efficiency principles, and in that sense as neutral. The emergence and consolidation of corporate power appears, then, as an inevitable form that economic growth takes under these ordering principles. The impressive engineering and architectural output . . . are a physical embodiment of these principles.
>
> (Sassen, 1996: 192)

Sassen is writing of North American cities in which corporate power literally towers over the street, but the model can be applied to projects such as Cardiff Bay, where contemporary art becomes the sign of 'success', whilst obliterating the political and ecological aspects of development. So, during the abolition of history, birds (who neither vote, pay taxes nor consume) are cleared out so that developers may inscribe on the site (in turn inscribing the development on the city) the kind of waterside vistas which are already the stereotypical images of their publicity brochures.

The policy for art in Cardiff Bay illustrates that ethics and opportunism do not mix. This is, in a way, to unfairly overlook the considerable body of art commissioned for the development sites, but that is not the point. The model of policy provided by the *Strategy* is in keeping with the dominant ethos of short-term gain in Thatcher's Britain. Would the same team of consultants who wrote the report for Cardiff Bay develop an art programme for a nuclear reprocessing plant, or a concentration camp, and if so would any quality of art justify the activity it masked?

PUBLIC ART IN THE PUBLIC AND PRIVATE SECTORS

Birmingham

Birmingham City Council has integrated art and urban design in the re-development of a series of public spaces – a pedestrianised New Street, Chamberlain Square, Centenary Square and Victoria Square – linking the new International Convention Centre (ICC), Repertory Theatre, public library and City Museum and Art Gallery, as well as international hotels. These schemes constitute a new city centre 'framed' by distinctive Victorian civic architecture; the ICC is perhaps the organising principle of the development.

The city's public art programme is embedded in a wider drive for economic regeneration, and its strategy has been to promote the city internationally through the image of its central district and a cultural profile which includes its Symphony Orchestra. Selwood notes a tradition of public art in Birmingham from nineteenth-century monuments and architectural embellishment to modern sculpture, and reviews the commissioning processes, which varied considerably between the two squares and were not without conflict (Selwood, 1995: section 5). A sponsorship brochure issued by the City Council states: 'The City of Birmingham recognises the value of investing in art to build for a successful future' and 'industry is recognising that there are both long- and short-term benefits to be gained from investing in the arts'[21] (Birmingham City Council, n.d.). Birmingham's Unitary Development Plan draws attention to the Percent for Art policy adopted for the ICC,[22] and its own policy of commissioning artists to work with design teams on environmental enhancement schemes.

Projects in Birmingham combine two different but not incompatible aesthetic strategies: first the integration of art in urban design, in Tess Jaray's designs for paving and street furniture in Centenary Square; second the commissioning of contemporary art for sites within the redevelopment, including, in Centenary Square: a fountain, *Spirit of Enterprise*, by Tom Lomax; Raymond Mason's *Forward*, David Patten's *Monument to John Baskerville*; a mural designed by Deanna Petherbridge inside the ICC, and Ron Hasleden's neon piece *Birdlife* at its entrance;[23] in Chamberlain Square, Siobhan Coppinger's bronze of Thomas Attwood MP; and in Victoria Square Dhruva Mistry's pair of stone sphinxes *Guardians*, his sculptures *The River* and *Youth* in the fountain, and Antony Gormley's *Vision* (Figure 32).[24]

In Centenary Square, Mason's stereotyped images of 'the crowd' hurl their rhetoric into the air, whilst Jaray's paving and street furniture suggest a very different aesthetic. Whilst Centenary Square exhibits the work of several artists in competing styles, Victoria Square has the unifying aspect of one person's work, though its design was contested within the authority;[25] Mistry's eclectic works reference Baroque piazze, as well as Greek and Indian art – Lynne Green argues

FIGURE 32 Antony Gormley, *Vision*, Victoria Square, Birmingham

that Mistry's work creates a public through its connectedness to 'everyday life': 'This has much to do with the sculptural tradition of India, where art and its embodiment of material and metaphysical knowledge is understood by all' (Green, 1993: 40) – yet there are also hints of Empire in these imposing, colourless figures and the overwhelmingly hierarchic design of the 'stage-set' square. Both squares are well-used and are thoroughfares between a number of locations; seating, planting and (in Victoria Square) water are included in the designs. Postcards of the art in Centenary Square can now be bought at New Street Station, indicating its location on a tourist 'map'. In one way, the elements compete like art styles in the different rooms of the Royal Academy summer exhibition; in another they are united by a common intention, to re-inscribe the city in visual memory, and in this the distinctiveness of each image is what matters.

Public art in Birmingham has a key role in the manufacture of 'place'; Zukin writes (on New York):

> By the 1990s, it is understood that making a place for art in the city goes along with establishing a place identity for the city as a whole. No matter how restricted the definition of art that is implied, or how few artists are included, or how little the benefits extend to other social groups outside certain segments of the middle class, the visibility and viability of a city's symbolic economy plays an important role in the creation of place.
>
> (Zukin, 1996: 45)

The definition of place, which implies boundaries between 'here' and 'not-here', boundaries which shift and are perceived according to certain values, is, in sociology or cultural geography, problematic;[26] for the advocates of conventional public art it is seen as a physical site.[27] In Birmingham, 'place-making' has been interpreted as urban design allied with public art, integrating the decorative and functional in Jaray's work, but otherwise retaining separate identities for art and design within a common professional ideology. The fact that design issues were contested is part of that professional framework and its rivalries, not a fracturing of it; what unites the whole programme is that it serves to give a public face to development designed in a framework of corporate values in which the ICC is the organising principle not only spatially (the hub of the site), but also conceptually (the core value is commercial development). This is an approach which privileges the visual in constituting the city; whilst the venues around Centenary Square are for the performing arts, the dominant image is the creation of grand vistas.

A *central cultural district?*

The encasement of the corporate within the public raises questions about the designation of the spaces, which seem open to use and sight, and clearly are public space (in public ownership) not the pseudo public space of shopping malls or the Broadgate development in London. Yet the same spaces, which are an image of the city for distanced consumption, represent a public sector investment intended to attract private sector economic growth, and are remote from many publics and parts of the city. Selwood reports that the strategy began with investment in the National Exhibition Centre which, since its opening in 1976, has attracted new business (and businesses) to Birmingham, and is based on the model of cities such as Baltimore (Selwood, 1995). She also cites a report by Patrick Loftman and Brendan Nevin of the University of Central England which states: 'Advocates of this kind of model of regeneration justify massive public sector expenditure on the development of prestige projects on the basis that they are directly linked to the well-being of all city residents' (Loftman and Nevin, 1992: 6); their report challenged the economic success of the schemes and was rubbished in the City Council. Sir Nicholas Goodison of TSB cited the cultural programme as one of the reasons for his company relocating to the city (Selwood, 1995) which recalls Myerscough's findings, bearing in mind that the 'cultural programme' includes the ICC and Repertory Theatre, as well as museums and conservation areas. A question remains, however, as to whether the employment created by the develop-ment is only part-time and low-paid and, crucially, whether what is perceived as a civic centre transformed into a 'central cultural district' appealing to a middle-class public, is actually a 'central business district' on the Burgess concentric ring model, appealing to corporate interests to the exclusion of disadvantaged groups and those living in peripheral urban areas.

The re-inscription of the city, then, excludes difference. Yet Sassen writes: 'A large city is a space of difference', citing Sennett's *Conscience of the Eye* (Sassen, 1996: 41). Her concern is with cultural and ethnic difference, and the uncovering (to reclaim into value) of marginalised uses of urban space, in other words the production of spaces conducive to mixed uses by mixed publics. The centre of Birmingham is somehow, in its appeal to corporate utopianism, too nice to do that.

Swindon

A different strategy, still linked to inward investment in a town where new, mainly insurance and electronic, industries have replaced railway engineering, has been developed by Thamesdown Borough Council, in Swindon. Thamesdown, which has supported murals and public sculpture since 1975, the same year in which the first gable-end murals were painted in Glasgow, and has a dedicated public art post at a senior grade within the authority, has prioritised residential areas as well as town centre sites for commissions, often through artist's residencies which include workshops and regulated public access to the artist at work. A sculpture commission by Hideo Furuta in 1986 was particularly successful, through the artist's communication skills and working methods, in progressively gaining the confidence of the public of a new housing area, and their elected representatives (Miles, 1989: 155). One of the town's commissions, *The Great Blondinis* (Figure 33) by John Clinch, in the Brunel Shopping Centre, whilst the kind of populist art loathed by the modernist establishment, was a collaboration between the artist and craftsmen from the rail engineering works, and stands in local memory as a monument to their skills.

Broadgate

If Thamesdown is a model of dispersed commissioning, and Birmingham of centralised commissioning, comparison with the Broadgate Business Park demonstrates the closeness of the Birmingham model to the corporate. Broadgate, adjacent to London's Liverpool Street Station (a private sector initiative of 3.5 million square feet of offices, the station's Victorian structure restored within the development), in which thirty or so artworks were commissioned at a cost of £3.5 million,[28] is a corporate enclave; but whilst in Birmingham public space has been preserved, the pseudo-public spaces of Broadgate are internalised. The courtyards, though well-used in Spring and Summer lunch breaks and sometimes sites for free entertainment, are interior to the development and do not invite casual entry or open onto public streets and, like the entries from surrounding streets, are watched over by closed circuit television cameras.

Broadgate is an enlightened 'corporate fortress', from within which the rest of the city is invisible. A *Visitors' Guide* states that 'Using sophisticated management techniques and innovative construction methods, the buildings were constructed to a quality rarely found in British commercial buildings' and the list of artists whose works have been purchased reads like a 'blue-chip gallery' stable.[29] Botero's *Venus* proclaims its unreconstructed appeal to the masculine gaze, described in the *Visitors' Guide* as 'voluptuous . . . a connection with the primitive' (Broadgate, n.d.).

BATTERY PARK CITY

Battery Park City provided a 'world-class' model of development on a 92-acre strip of land reclaimed from the Hudson River[30] near the southern tip of Manhattan, with 16,000 units of housing and a commercial centre, and from which there is a view of the Statue of Liberty. Savitch describes it as 'designed to be a high quality, high-rent neighbourhood for the elite of post-industrial New York' (Savitch, 1988: 51). Battery Park City is celebrated as a case of artist–architect collaboration – Cesar Pelli with Scott Burton and Siah Armajani at North Cove, for example, and at South Cove, at the far end from Pelli's tower, Mary

FIGURE 33 John Clinch, *The Great Blondinis*, Swindon

FIGURE 34 Battery Park City – a waterside landscape and a viewing platform designed by Mary Miss

Miss with landscape architect Susan Child and architect Stanton Eckstut – which produced well-designed public spaces. In particular, Ned Smyth's *Upper Room*, a fusion of art and architecture in one of the small plazas, referencing the 'upper room' of Leonardo's *Last Supper*, is used by families who picnic on its table. But art in Battery Park City can be critiqued in different ways: as art-and-architecture, or as an element in the continuing and divisive configuration of New York; two models of criticism, from John Beardsley and Rosalyn Deutsche, demonstrate this.

Sympathetic description

Beardsley writes of South Cove: 'While the landscape recalls a natural inlet, other features of the design are associated with the history of the waterfront, especially the wooden piles and decking' (Beardsley, 1989: 152–3), suggesting a kind of naturalisation. His account reflects modernist art criticism – Battery Park City represents the possibilities of collaboration, which might be taken as a meeting of professional ideologies on Brighton's 'siamese twins' model noted in Chapter 4. Beardsley's description concludes that the landscape is 'exemplary', with a coherence suggesting the feeling of a neighbourhood; this is produced by the architecture and design of public spaces 'which knits the string of parks and plazas together along the esplanade'; he adds 'as a form of civic expression, public art . . . has emphatically improved. There is little doubt in my mind that this art has had a beneficial impact on the public space' (Beardsley, 1989: 155). Beardsley sees the scheme in terms of a relation of spaces, a formal and mechanistic approach which is value-free, although he is worried it neutralises the art.

Critical enquiry

Rosalyn Deutsche takes a different approach, situated in relation to dialectical materialism and post-structuralism: for her, 'Spatial forms are social structures' (Deutsche, 1991a: 160)[31] and art in Battery Park City 'violently fractures the social picture . . . conceals domination . . . rejects time' and tries to integrate Battery Park City with a New York which it re-figures 'ghettoized and exclusionary' (Deutsche, 1991a: 202). Deutsche's point of departure is the image of homeless people, their 'plight' seen by the dominant culture as either an inevitable aspect of metropolitan life, or as individual misfortune: 'To legitimate the city, this response delegitimates the homeless' (Deutsche, 1991a: 158). Hence homelessness is either 'natural', in a legacy of Burgess' natural science idea of zones of transition as the city receives waves of 'invasions', or else it is self-inflicted. Deutsche cites Mayor Koch's remarks criticising a court decision to squash a city anti-loitering law designed to remove the homeless from Grand Central Station, and points to the way the assignment of simple function to place, whilst rational, 'is to claim that the city itself speaks', or that individual locations have an inherent meaning 'determined by the imperative to fulfil needs that are supposed to be natural, simply practical' (Deutsche, 1991a: 159), denying the social production of spatial meaning, and the complex mechanisms by which homelessness (eviction) is created by development and gentrification. Deutsche cites Lefebvre's view of space as ideological because it reproduces existing social relations and Manuel Castells' description of the city as a site of conflict and resistance to domination, summarising the process as 'the inscription of political battles in space' (Deutsche, 1991a: 161); she then contrasts high-rent office towers, luxury condominiums, corporate headquarters with the predicament of homeless (or evicted) people, following the gentrification of their neighbourhoods, and argues that the divisive and repressive social policies which facilitate development justify themselves through art, as beauty, though artists could subvert the process:

> As a practice within the built environment, public art participates in the production of meanings, uses, and forms for the city. In this capacity, it can help secure consent to redevelopment and to the restructuring that make up the historical form of late capitalist urbanization. But like other institutions . . . it can also question and resist those operations, revealing the supposed contradictions of the urban process.
>
> (Deutsche, 1991a: 164)

which seems similar to Gramsci's model of hegemony and anti-hegemony[32] – a particular economic interest using the institutional structure of the state, in this case art, to gain an acceptance of its programme, and the possibility to resist through the same structures.

Deutsche sets out the contradictions, as in the emphasis on public space in privatising development – 'the provision of space for "the public" testifies . . . to the wholesale withdrawal of space from social control' (Deutsche, 1991a: 165), producing 'disciplinary technologies in Foucault's sense insofar as they attempt to pattern space so that docile and useful bodies are created by and deployed within it' (Deutsche, 1991a: 187) and resulting in what Alexander Kluge and Oskar Negt have theorised as a 'pseudo public sphere' (Kluge, 1981–2) in which private interests increasingly control the public domain and commodify all aspects of daily life, and in opposition to which they propose an 'oppositional public sphere' described by Deutsche as an arena of political consciousness, as illustrated by the project *If You Lived Here* with its focus on homelessness and representation (Rosler, 1991).

Deutsche goes on to discuss art integrated with urban design – such as Scott Burton's seats – which is 'useful in the reductive sense of fulfilling "essential" human and social needs' (Deutsche, 1991a: 173). Against a tendency which, in the 1980s, had distinguished two kinds of public art – as: conventional (a re-run of the monument, a new kind of 'canon in the park' or simply an extension of the market for sculpture); and integrated or useful – Deutsche conflates both categories as affirmative of the dominant ideology, and establishes a new alternative as critical public art which interrogates its location in the public realm.

Deutsche goes to some length to demonstrate the pivotal role of art in Battery Park City when the lower-income housing component was dropped from the scheme (Deutsche, 1991a: 191–202); fiscal pressure was used to make radical changes in the investment appeal of the plan, the land was transferred to an Urban Development Corporation, and plazas were incorporated to provide an up-market feel as the financial component was re-located to the centre of the development. Parks and plazas were introduced at the same point as tax breaks and other incentives to make the scheme financially attractive 'by creating parks and other amenities to convert the area into an elite district as quickly as possible' (Deutsche, 1991a: 194). The alliance of art and the exclusivity of Battery Park City has, then, more than a metaphorical aspect. The kind of modernist art with which the scheme is enriched is trans-historical, and thus has no space of critical engagement; the co-option of art to beautification parallels art history's aesthetic approach to the city, which Deutsche sets out in 'Alternative Space':

> All connections between art and the city drawn by aestheticist tendencies within art history are, in the end, articulated as a single relationship: timeless and spaceless works of art ultimately transcend the very urban conditions that purportedly 'influenced' them.
>
> (Deutsche, 1991b: 46)

And in so doing they abdicate responsibility.

London Docklands

Like Battery Park City, London Docklands is a new city, though it is built on a site with a long history; there are other differences, too: it is more mono-functional (a corporate zone of 12 million square feet of office space in Canary Wharf alone, near to, but not integrated with, enclaves of high-cost housing) and its artworks are more conventional. Jon Bird, in his essay 'Dystopia on the Thames', contrasts the harmony and coherence of the aesthetic view presented by the developers and the London Docklands Development Corporation (LDDC) with the alienating experience of visiting Docklands, and with the view of the committee constituted of locally elected representatives (the Docklands Consultative Committee) that 'the real beneficiaries were service-sector industries and employees, with only a minimal proportion of jobs being made available to local residents, mostly on government training schemes and temporary and part-time servicing contracts' (Bird *et al.*, 1993: 126–7). Ambrose reports that much land in the borough of Tower Hamlets was appropriated by the LDDC to help promote speculative, up-market development (Ambrose, 1994: 24) and cites the accounts of a local GP, David Widgery, who says of the development period, the 1980s: 'one became hardened to the sight of hunched and dishevelled human bundles in doorways, disused buildings, even skips' (Widgery, 1991: 120, cited in Ambrose, 1994: 182). Widgery gives several succinct images, for example: 'Canary Wharf remains curiously alien, an attempt to parachute into the heart of the once industrial East End an identikit North American financial district' and 'Cesar Pelli the architect of the fat Canary tells us that "A skyscraper recognises that by virtue of its height it has acquired civic responsibilities" . . . Now that would be an interesting idea' (Widgery, 1991: 41 and 164, cited in Ambrose, 1994: 182–4).

The LDDC has an art programme based on a report in 1989 by Comedia, *Creating a Real City: an arts action programme for London Docklands*, including links with local galleries, encouragement of artists' studio provision, and public art. There is a greater emphasis on a mix of arts activities than in the *Strategy* for Cardiff Bay; but there is also conventional public art. Some sculptures, such as Wendy Taylor's shining steel *The Spirit of Free Enterprise* at West India Dock, or Bruce McLean's sculptural railings at Canary Wharf, were already in place before the launch of the art programme; the brochure states:

> The built and natural environments in London Docklands are rapidly chang-
> ing, with new buildings, roads and railways, alongside new open spaces,
> parks and landscaping schemes. Artists and craftspeople can work with
> architects, engineers and developers to great effect, enriching and humanis-
> ing the environment through public sculpture, floorscapes, cladding, murals,
> architectural carving, land art, wrought ironwork and street furniture.
>
> (LDDC, 1991)

The social conflicts of Docklands are known through the Docklands Community Poster Project (Bird *et al.*, 1993: 136–49), but the model of commissioning sculptures is similar to that of Battery Park City: modernist art and architecture are utilised by corporate interests to persuade a notional public, addressed through media images, that the scheme is 'good'; whilst the aesthetic gloss conceals contra-dictions between the responsibilities of public authorities and their collusion in policies which are anti-social but in the interests of corporate investors. Are there alternatives to this complicity? The poster project is a model of resistance, but are there possibilities for collaboration with local communities if development is more in their interest than at Docklands?

SUNDERLAND

If Battery Park City, Canary Wharf and Cardiff Bay can be compared as develop-ment which is anti-social or anti-ecological, but legitimised by art, the approach taken in the redevelopment zone at Sunderland is an alternative, although the planning authority is a Development Corporation and the site, from which all traces of the ship-building industry have been erased, represents another abolition of history, and with it an abolition of employment. Tyne & Wear Development Corporation (TWDC) is responsible for regenerating 2,428 hectares of land along

FIGURE 35
Children at *The Red House,* Sunderland

30 miles of the Tyne and Wear, including the St Peter's Riverside area of Sunderland where the art project began in 1989 with a feasibility study by the Artists Agency. The study found that local people, in Monkwearmouth and Roker, saw potential in using the arts to focus and promote community identities and to influence the development of the area (Artists Agency, 1996: 4).

Rather than approach the TWDC and private sector developers as patrons of conventional public art, the Agency began by bringing together a group of local people and community groups, schools and churches, with a local artist, Colin Wilbourne, to further research the possibilities:

> The once-familiar tendency to parachute artists into run-down residential or commercial areas to 'cheer up the place a bit' is increasingly rare, but the legacy of those ill-conceived adventures is still evident in housing estates and shopping centres and asks the following questions as ground-work: Who uses the space? What's its history? What contribution might art make?
>
> (Artists Agency, 1996: 5)

Whilst one aim of the residency is to create sculptures carved from the sandstone remaining from demolition of previous buildings, to lend 'visual cohesion' to the site on the usual pattern – no-one seems to propose art as adding difference, or what Sennett terms 'disorder' (Sennett, 1996) to a site – the social aspects and processes of engagement seem equally important. What is not intended is to make the site either a museum or a memorial to ship building; at an early meeting with the steering group of local people, the artist was asked not to refer too directly to the lost industry, rather as if it might be still a private grief.

After seven years, at the time of writing, the artist is still on-site, and has been joined by three others; assistant sculptor Karl Fisher, writer Chaz Brenchley and blacksmith Craig Knowles. The difference between this project and standard public art solutions is in five particular factors: the involvement of the communities and a local arts organisation (the Artists Agency) through a steering group which considers all matters of public art in the development; the setting up of a long-term residency (beginning as a research project amongst the local communities, and including a training function) by a local sculptor; the extensive development of workshops and other outreach elements in the project; the low-key approach taken in making artworks which are part of the site and in appointing a regional rather than international artist; and the consortium of development partners with a socially geared profile. Housing provision and accommodation for the University of Sunderland were key elements of the scheme, and the Northern Rock Building Society one of the funding partners for the artist's residency. These factors are inter-related but in this discussion the relation to local communities is perhaps the most important, and is interpreted not as simply the manufacture of 'place' for which local people are a passive public, nor a colonisation of local history (although history is involved, for instance the Anglo-Saxon illuminated manu-scripts made nearby in the ninth century, referenced in a sculpture *Pathways of Knowledge*), but as the accountability of the artist to a representative group of local people. This is a social rather than geographical or aesthetic definition of locality to which the publicity material refers.[33] From these communities the steering group is derived, meeting quarterly. This approach has been possible because St Peter's Riverside is not a central business district masquerading as a cultural or civic centre, but is a former industrial area which has a long history as the main site of employment, reconstituted as a housing area. In a way it is a journey from one sort of margin to another, but in process, and in part through the art project, reclaiming a public domain. It therefore does not exclude the publics in whose midst and with whose participation it has grown. If anything, the art plays on marginality, constructing 'places' in incidental locations – a look-out point, a flight of steps with a carved pattern and pair of stone shoes, a 'ruined' house.

Memories of place are utilised in *The Red House*, made from red sandstone from the Queen Alexandra Bridge and sited on what is called the Barbary Coast, on which houses pre-dated the ship building yards, and have returned: a few fragments of a house wall, a door, chairs and a carpet, a coat on the door. *The Red House* is filled with things, some appropriate to the reminiscences of elderly people, and based on the 'tray game' in which a player is shown a tray with objects which is then taken away, the player trying to remember as many object as possible; but the occupants have left – migrated? died? The work functions as a monument to everyday life.[34]

Evaluation

The project has been monitored regularly and evaluation reports produced. One stated aim is 'creating a sense of ownership of the artworks by local people', another 'Offering access to the public to view works in progress and access to the artists through exhibitions, workshops and public participation' (Pym, 1995);[35] evaluation is reliant on anecdotal evidence,[36] but the low incidence of vandalism is also evidence of a sense of ownership.

Deutsche's critique might be applicable here; the redevelopment is located in the abolition of an industry, and it could be argued that the sculptures naturalise it by seeming to be part of the original substance of the site. On the other hand, the point of departure and process are different from those of art in corporate developments. Perhaps there is a nostalgia in some of the work, or perhaps it is a local history engendered by personal memories of place. It remains, within the terms of art as a material process rather than a purely social process, a model of community involvement which declines to offer corporate baubles or controlling utopias.

URBAN DEVELOPMENT AS HEGEMONY

With the exception of Sunderland and Thamsedown, the cases discussed above indicate a complicit role for art in development, concealing its social contradictions

FIGURE 36 *The Red House,* Sunderland – a coat behind the door

FIGURE 37 A
relief depicting
Edward II by
children in Swansea

in aesthetic gloss, at Cardiff Bay concealing ecological vandalism. This is not to argue against the collaboration of artists in the design of public space, and not to argue against the provision of public spaces in development. It is to problematise the complicity of public art with corporate development and the structures of money and power on which it is founded. Developers do not develop in order to construct the 'city beautiful', they construct the city beautiful in order to conceal the incompatibility of their development with a free society.

Art in urban development is a case of hegemony, in which the status quo, that is, freedom for capital to increase and the unfreedom of the majority population to determine the conceptualisation of the city, is preserved. The mechanism appears to be: the economic interests of corporations, who finance, manage and occupy development, are advanced in ways involving both public sector subsidy, and a public presentation of development as beneficial in which the state acts, through planning de-regulation, as facilitator; society's institutions – architecture, landscape design and public art – dress the development with aesthetic quality. The resemblance of development to art thus locates it in a domain which is outside political contention and contained in a territory of expertise which inhibits the comment of non-experts (whose eyes are insufficiently 'educated' to 'appreciate' it), within a utopian fantasy which is a means of control; this is sustained by the naturalisation of social problems, or the classification of the socially disadvantaged (including those disadvantaged by development) as dirt. The glittering glass façades, the sparkling water, conceal the disempowerment of urban dwellers to occupy the space of determining what the city is, produce the ejection of communities through gentrification, and disaffection of communities in adjacent areas for whom the benefits remain 'over the fence'. But perhaps this is exactly the kind of function for which modernist art is suited.

6

ART AND METROPOLITAN
PUBLIC TRANSPORT

•

INTRODUCTION

During the 1980s, public transport became a growth area for public art, almost becoming a sub-specialism,[1] as artists were employed to make work for stations and collaborate in the design of transport facilities.[2] In one way this continued the tradition of the London Underground, which had employed artists to design posters for its stations since Edwardian times, and the New York Subway, in which stations built at the beginning of the century were embellished with decorative tiles; but it also created new kinds of opportunity for artists to engage in collaborative processes of design. This development took place in context of a wider involvement of artists in the design of public spaces, as in several projects in Seattle,[3] and at Battery Park City. But whilst art in urban development may, as argued in the previous chapter, be complicit in abjection, public transport is seen as a social 'good' available to most members of an urban society, decreasing pollution and levels of energy use, thus contributing to urban sustainability.[4]

The role of art might, then, be to encourage the use of public transport by obvious strategies such as making it (visually) more attractive or locally distinctive, thereby contributing to a positive public image of a transport system.[5] It is hard, and would be futile, to argue against this;[6] but it remains a territory bounded by the issues of design and decoration, and which more often relates, as site-specific art, to the static spaces of stations than the ephemeral experiences of movement. It also tends, again in keeping with the conventions of public art, to address the site as a physical rather than social space, and does not approach the public issues of transport policy. Elsewhere, establishing a new kind of single-issue politics, people from diverse backgrounds and allegiances are chaining themselves to trees on the routes of new roads, and in one case using the voluminous skirts of carnival figures as a protective screen whilst drilling holes in a freeway and planting trees.[7]

This chapter takes a selective view of art in metropolitan transport, based mainly on the New York Subway and London Underground, as two of the most extensive systems with the longest records of commissioning art and craft. There

are many other examples of art in metro systems, including the Detroit People
Mover, Boston T, Seattle Metro Bus Tunnel and the metros of Baltimore,
Newcastle, Paris, Amsterdam, Brussels and Stockholm, some of which are
described in other publications;[8] being specific to metropolitan transport, this
chapter does not cover, except in passing, art in the national network of British
Rail, nor the ubiquitous 'airport art' of international terminal buildings.[9] It begins
by briefly sketching as a context the impact of metro systems on cities, describes
a number of cases of art in the London Underground and New York Subway, and
then compares the 'progressive' messages of art in London Underground's unified
design programme of the 1920s and 1930s with the more nostalgic content of
some of its recent commissions for art in station refurbishment. This leads into
a discussion of the wider issues of which publics are addressed by art in public
transport, and whether it is a field in which critical messages can be conveyed.

METRO ART AND CITIES

The introduction of public urban transport in the late nineteenth century was a
factor in the zoning of cities; in enabling people to live on the city's edge but
travel easily and quickly to shops and offices in the centre, it contributed to subur-
banisation. In London, the emerging lower middle class were (with the introduction of building society loans) able to move away from pockets of dense poverty
in the 'East End' to outer areas offering new housing, travelling to and from work
by underground train; in the 1930s, the extension of the network created access
to the 'countryside', most of which duly became suburbanised. Sennett, relating
city form to the circulation of blood in the human body,[10] writes: 'During the
day, the human blood of the city flowed below ground into the heart; at night,
these subterranean channels became veins emptying the mass out of the center'
(Sennett, 1994: 338). The decreased diversity of use in the centre is a factor identified by Jacobs as contributing to city decline (Jacobs, 1961), but the mass transport which enabled this zoning in the early part of the century is now seen as
socially beneficial, vital once the spread of the city into a metropolitan region has
taken place. Most of the early systems were regarded as having the same importance
as civic buildings, and were, like city halls, suitably embellished.

   The metro entrances designed by Hector Guimard between 1900 and
1913 have become emblems of the Paris of the belle époque;[11] they suggest
cosmopolitan ease, using the design style favoured by the progressive taste of the
time – Art Nouveau – in which form and decoration are fused. The metro in
Moscow, with its lavish depiction of the soldiers, workers and peasants of the new
society, could be described as a 'people's palace',[12] and perhaps the (much later)
Washington Metro responds through its functionalism, expressing modernity
through technical achievement in the scale of its engineering rather than pictorial

representation or decorative finishes. The proprietors of early systems, such as the New York Subway, in which $500,000 was budgeted for decorative features in 1900,[13] saw design as a denotation of status, following the precedents of nineteenth-century rail termini. But whilst the rail termini were built like grand gateways facing back to the city, their allegorical sculptures co-opting the new enterprise to the myths of the nation state and its hierarchies, metro travel in London, Chicago and New York was one-class; images which signified aristocratic or bourgeois privilege, as in the classical allegories representing provinces and destinations which decorate many façades, were replaced by abstract and consciously 'modern' forms of decoration in the tiled walls of platforms and entrances. This modernity of everyday life is epitomised by the sans-serif type-face created by Edward Johnson for London Underground in 1916, commissioned by Frank Pick, who describes it as having 'the bold simplicity of the authentic lettering of the finest periods [and] belonging unmistakeably to the twentieth century' (Pick cited in Green, 1990: 10). The Underground, like the Subway and Paris Métro, was, then, a signifier of modernity, with all its optimism and utopianism.

ART IN THE NEW YORK SUBWAY

Art in the refurbishment of the New York Subway follows the precedents set by the builders of the original IRT, but for more pragmatic reasons. Graffiti and crime (and the identification of graffiti with crime) had by the 1970s created a negative identity so that middle-class commuters drifted away from the Subway to road transport, mainly their own cars; Sennett notes the scale of graffiti and its signification of an ever-present underclass perceived as menacing mainstream society, writing that 'transgression and indifference to others appeared joined in these simple smears of self . . . treated from the first as a crime' (Sennett, [1990] 1992: 205–6). In response, the New York Metropolitan Transit Authority (MTA) agreed a strategy in 1981 for a $16.3 million programme of system refurbishment as a way to make change visible. The MTA followed the lead of several other systems[14] in commissioning art within its programme of station upgrading, at about the same time that London Underground initiated collaborations between artists and architects in refurbishment of its stations. Art was, however, only one element in the MTA's strategy, which also included safe waiting areas, closed-circuit television cameras, enhanced levels of station cleaning and better lighting, alongside efforts at better punctuality. One of the most important aspects of the attempt to renew public confidence was constant removal of graffiti and introduction of new, air-conditioned cars. When graffiti-free trains arrived at stations on time, users began to return to the system but also to notice the condition of stations (Feuer, 1993: 141), strengthening the case for art.

Under a Percent for Art policy, 1 per cent of the capital costs of station upgrading was set aside for art, and in 1985 an Arts for Transit office was established to co-ordinate the programme with a brief to install permanent and temporary visual art projects, and produce dance, music and other types of performances, working with professional artists, art students, graphic designers, community groups and cultural institutions (Feuer, 1993: 141).

By 1987, six Manhattan Subway stations had benefited, and four in Queens and Brooklyn. Between 1988 and 1995, a further sixty permanent works were installed, including work by Maya Lin (Penn Station), Andrew Leicester (Penn Station), Milton Glaser (Astor Place) and Nancy Holt (Broadway-Nassau); some works, such as Laura Bradley's *City Suite* (96th Street), a set of designs in marble mosaic, ceramic tile and floor tiles, echoed the colours and motifs of the original IRT construction, elements of which are preserved within the refurbishment of many stations; others, such as the set of four mosaic murals by Lee Brozgold and students of Public School 41 (Christopher Street) depict local history and street life above-ground in an effort to gain community ownership for the station. In some cases art aids way-finding – at Kings Highway (Brooklyn), for example, Rhoda Andors created porcelain enamel murals depicting symbolic figures with hieroglyph-like signs above and beneath; the signs on the Coney Island side (where people walk towards trains for the beach) include fish, sea and boats, whilst those on the Manhattan side relate to work.[15]

Most of the commissions were for art added to the station environment, but two artists produced designs for station ironwork, the first being Valerie Jaudon, who undertook a pilot project in 1988 to design new railings (to separate the open area from that for which token entry is required) at 23rd/Lexington, titled *Long Division* (Figure 38). Despite a lengthy process of design modification and the increased cost of artist-designed railings, a system-wide replacement of barriers was introduced, using two modular designs by Laura Bradley, one based on waves for stations designed in the 1920s and '30s, the other on cut squares and circles for the older stations in which the medallion motif is used in the original mosaics. The constraints included that bars should be no more than five inches apart (to avoid children's heads being stuck between them), that the modules should fit 700 sites, and that they should be in stainless steel. The height of the barriers was increased in the process, to prevent fare evasion, but the aesthetic impact appears to have distracted attention from this security measure (FTA, 1996b: 33).

A recent addition to the subway art programme, based on a model used by London Underground for nearly a century, is the commissioning of four artists each year to design posters representing destinations within the system (which can be reached 'for the price of a token'). The localities are chosen by the artists, who are selected by a panel of arts professionals; the print run is 4,000, with

FIGURE 38 Valerie Jaudon, *Long Division*, railings designed for the New York Subway (Photo: Arts for Transit)

1,000 copies reserved for schools, MTA offices and sales through the Transit Museum, and the rest located in advertising spaces which have not been sold to advertisers.

ART IN THE LONDON UNDERGROUND

In the 1980s, artists were employed by London Underground to work with architects and engineers on the design of platform and circulation areas in a programme of major refurbishment, termed 'Changing Stations'. Although the system did not face the same loss of confidence as the subway, and its use increased under the Greater London Council's 'cheap fares' policy, a business case for visual improvements was established from surveys of passenger preference in the late 1970s, and a £65 million programme was authorised by the GLC in 1981. The involvement of artists was seen as an attempt to revive the tradition established by Pick, within a new design strategy.[16]

Setting up artist–architect collaborations, however, was not what Pick had done – his design style was produced by designers, not artists, and the role of artists was specific to creating images for posters. The origin of the 1980s strategy is perhaps more, then, in the successful advocacy of bodies such as the Arts Council, who financially supported the first design competition for Holborn Station.

The design concept for the refurbished stations was, in any case, defined before artists were approached; in 1979 a life-size model was constructed on a disused platform at Aldwych, to demonstrate and test the new concept. A kit of parts (mainly coloured tubular elements) was used to establish, through the colour red (taken from the Underground map), a line identity for the Central Line in the architectural detailing of the platform – something which would be immediately recognised by travellers and assist people in finding their way in busy interchange stations. Between 1980 and 1982, schemes were devised for Holborn, Queensway and Oval stations but not built; at Holborn a design competition was held for the vaulted ceilings of the escalator shafts. The architect's note[17] outlined the design concept and its three levels of visual identity:

1 Underground system identity
2 Line identity
3 Station identity.

The strategy was: to ensure a commonality of key elements to express the system identity, such as the well-known logo, the typeface and the standard location of signs on platforms; then to enable passengers to recognise lines by colours from the map, and stations by images, each having a characteristic design related to its surface location. The policy identified the interests of three groups: passengers, for whom it offered 'pleasure' and a 'sense of identification with part of what is at the moment an unresponsive and alienating system'; the operator, who required 'increased marketing potential' and 'improved personnel conditions'; and the community as a whole, who might be more encouraged to use the system (LUL, 1993: 8).

The competition for Holborn sought 'visual interest' and a re-examination of ways in which it might be achieved by artists, who were asked to accommodate the safety and lighting needs of the site (and reminded to buy a ticket when making site visits),[18] but whilst the competition produced several colourful proposals, none were thought practicable. Donald Hall, Director of Architecture and Design for London Transport, commented:

> The competition for Holborn fell flat, as competitions so often do, although I think it was our fault; we produced the wrong brief and the jury weren't at all concerned with the practicability of the schemes. I'm looking again for artists for Holborn, which we're thinking about as being the station for the British Museum. We try to establish a theme for a station.
>
> (Petherbridge, 1984: 24)

Later projects concentrated on the platform spaces rather than the escalator vaults, using a range of highly durable materials such as tiles and steel enamel. Some artists, such as Nicholas Munro at Oxford Circus (figures in a maze and snakes and ladders) and Michael Douglas at Baker Street (a head of Sherlock Holmes), designed repeat patterns for tiling. Others, such as Chris Tipping at Waterloo (timeless musicians and faery-folk) and Annabel Grey at Marble Arch (fantasy arches) and Finsbury Park (fantasy balloons), made individual designs for each tile or steel panel. In some cases, such as Green Park, London Underground's in-house architects designed tile patterns relating to the above ground station identity, or, as at Piccadilly Circus, using broad geometric motifs derived from the 1930s style, rather as Laura Bradley had done in the Subway, though abstract designs depart from the strategy of referencing above-ground locations.

The first pictorial scheme completed was Charing Cross, with images from the National Gallery reproduced on the Bakerloo line platforms and a depiction in black and white of the erection of medieval 'Eleanor' crosses by David Gentleman on the Northern Line. The best known scheme is Paolozzi's work for Tottenham Court Road, completed in 1986, also the most expensive, involving the hand-setting of mosaic tesserae; his designs evoke popular culture, picking up on the proximity above ground of the music publishing, film, and hi-fi retailing industries. In a few cases, artists working in a formalist style were commissioned; at Embankment (1987), Robyn Denny produced a composition of bars of saturated colour on a white ground, including a signature. Contrary to some expectations, the white expanses have not been covered in graffiti, although the occasional 'unofficial' signature mimics that of the artist. Other instances include Paul Huxley's blocks of colour at King's Cross (1987), and David Hamilton's and Robert Cooper's work for Euston, using abstracted elements of the heraldic arms of the Dukes of Grafton. The architects were not always happy in their collaboration with artists whose interests were in the aesthetics of modernism rather than the functions of transport, and a new design policy was adopted in 1990: 'to provide as its key objective a simpler background for passenger movement and information to deal with increasingly higher usage' (LUL, 1993: 11). Peter Dormer, who sees the 1980s programme as 'patching up with art' and visually confusing for passengers, writes of the 1990 policy:

> Now, London Underground is trying to make its stations an example of civilised utilitarian design . . . station designs by the award-winning international architect Santiago Calatrava for new open air stations are an example of [a] readoption of the art of architecture and design – holistic modernisation, not patching up.
>
> (Dormer, 1993b: 40)

and Wellington Reiter argues that although the involvement of artists in design processes has become a 'mantra' for public art advocacy, 'artists may find a place at the conference table, but in taking that seat they tacitly agree to work with

FIGURE 39 Eduardo Paolozzi, mosaics for Tottenham Court Road Underground station

FIGURE 40 Art from the National Gallery in the London Underground

FIGURE 41 Art from the National Gallery in the London Underground (detail)

fellow-designers, thus relinquishing the most fertile territory of the arts: critical commentary' (Reiter, 1995: 66).

MODERNITY AND NOSTALGIA

It is difficult to read a coherent message beyond a reflection of current styles in most art programmes in public transport, including the Changing Stations programme of London Underground in the 1980s, with its mix of minimalist abstraction, decorative pattern and figurative imagery.

Modernity

For a clearer statement of meaning it is necessary to look to London Underground's co-ordinated design style of the 1920s and 1930s, through which it proclaimed the optimism of the machine age in the visual language of international modernism; its extension into suburban areas produced a distinctive architectural style in stations such as Arnos Grove and Southgate, designed by Charles Holden, extending to details such as lighting columns for the platforms. The London Transport Passenger Board, established to manage the unified system for public benefit in 1933, was depicted by Lilian Dring in a triple poster as a new Mercury 'the modern god of transport' in a style reflecting Leger's use of local colour and tubular figures – his 'tubism'; the logo is Mercury's heart and the tube lines his arteries, a train becomes a snake towards which commuters walk and a sunburst behind his head confers mythic status (Green, 1990: 78). The design was not used for cost reasons, but Mercury did appear, with Persephone, in *The Underground Brings All Good Things Nearer* (1933), by Dora Batty typifying a whole genre of posters to encourage Londoners to explore the 'countryside'.

The inspiration behind the poster programme was Frank Pick, appointed in 1908 to the publicity office and remaining with London Underground until 1940.[19] Pick saw the entire visual appearance of the Underground as part of his brief, personally supervising it in all aspects. The detailing of carriage interiors, the textile designs for seating, lighting fixtures in carriages and on stations, the lettering style of notices, were co-ordinated within the London Underground style to represent the unification of a network which had originally been separate lines, and through which it promoted itself as a model of art in industry, and public service; the style was reflected in poster designs by artists such as E. McKnight Kauffer, Rex Whistler, Laura Knight, Graham Sutherland, Eric Ravilious, Edward Wadsworth, Paul Nash, Laszlo Moholy-Nagy and Man Ray. McKnight Kauffer, an American living in London, designed *Power* in 1931 – a muscular arm and fist emerging from the London Underground sign, surrounded by a mix of modern type styles, and his poster for the London Museum shows multiple imagery derived from his contact with Vorticism. *Speed* (1930) by Alan Rogers depicts machine-age man in the style of Leger and Le Corbusier, in black, red and blue, the figure abstracted amidst the stylised carriage and London Underground symbol. William Roberts, a member of the Vorticists, also contributed a poster, though not until 1951. McKnight Kauffer remains the designer who most sums up the approach; his *Winter Sales* of 1921 suggests awareness of Boccioni's *States of Mind* of 1911–12, itself an image of railways.

The message of the posters, like that of international modernism, is Progress; in the public spaces of the system, all were equal and all were looking forward to a better tomorrow. An intention to appeal to a broad public is demonstrated in texts such as 'The way for all'[20] and the relation of posters to popular culture – F. C. Herrick's design for a poster at the 1920 International Advertising Exhibition[21] depicted such advertising images as the HMV dog, Johnnie Walker, and the Bisto Kids. Whilst this range is as broad in its time as that of the 1980s refurbishment programme, the difference is the common aim established by Pick, who was adventurous, and who saw the commissioning of artists working in modernist styles as effective promotion of a modern system, but also restricted the presence of art to posters within a uniform station design.

Nostalgia

Today, after a gap in the 1970s when most promotion was contracted out to agencies, posters are again commissioned,[22] although their messages are in some cases subsumed in the agenda of tourism and heritage. Catherine Denver's *Explore London* of 1986,[23] for example, shows a female figure in a rather 1930s style of drawing, amidst London buildings set in a landscape devoid of people or traffic – a view of the city to which Londoners are not currently privileged.

In contrast to the optimism of the 1930s, some of the art in London Underground's 1980s station upgrading programme reflected a drift to heritage culture. A contemporary visual language is found in a minority of stations compared with the use of images suggesting a historical or fanciful past: of twenty-seven schemes completed by 1996, nine[24] have historical themes, four[25] use contemporary art styles, five[26] represent the new policy of cooler and more functional design, and the remainder are either restorations or simply decorative tile designs. Heritage is thus the largest single category, including a transposition of Sir Marc Brunel's drawings for the Great Western Railway, by David Hamilton, at Paddington (1986); references to the Natural History Museum, by Mary Woodin, at South Kensington (1988); images of hot air balloons, looking back to a supposed past of leisure and lightheartedness, by Annabel Grey at Finsbury Park (1986) and equally whimsical arches at Marble Arch (1986); and Chris Tipping's 'semi-allegorical images . . . richly decorated in the manner of medieval tapestry and manuscripts . . . evocative, transparent images like memories of things past'[27] at Waterloo (1988). At Holborn, its refurbishment finally completed in 1988, the platforms are decorated with vitreous enamel panels designed by Alan Drummond using photographic reproduction of objects from the nearby British Museum, a sort of burial to be explored by the passenger as archaeologist, perhaps. Tipping's figures seem lost in a dark blue sea of other-worldly time, and Drummond's hover in a chthonic darkness; both construct an aesthetic space which invites diversion from the cares of contemporary travel. This suggests not only that the optimism of the system in the 1930s has been lost in face of the pressures of declining investment, overcrowding, poor timekeeping, and worries about safety following a fire at King's Cross in which more than thirty people died, but also that, in answer to present realities, art offers, being what Marcuse termed 'affirmative culture',[28] an aesthetic escape route.

ART, TRANSPORT AND THE PUBLIC REALM

The messages conveyed by art in the London Underground establish the platform as an art space; at the same time, mass transit systems are locations of more informal mixing in urban society than almost anywhere else. The business executive may take the metro from an airport, as the fastest route to a central business district, and the stations are used by homeless people (because they are warm) and street musicians, as well as diverse urban publics constituted by gender, race, class, age and locality. This constructs stations as part of not just public space, but also of the public realm, in terms of the desired coherences of civic aspiration and the visible differences of urban cultures and sub-cultures. In 1996, the Federal Transit Authority published a policy that:

> Mass transit systems should be positive symbols for cities, attracting local riders, tourists, and the attention of decision makers for national and international events. Good design and art can improve the appearance and safety of a facility, give vibrancy to its public spaces, and make patrons feel welcome. Good design and art will also contribute to the goal that transit facilities help to create liveable cities.
>
> (FTA, Circular 9400.1A)

which sums up the case for art adopted by many transport authorities in the USA, and is compatible with both that for art in the New York Subway as part of a response to a declining public image in the 1970s, and that for art in the Seattle Metro Bus Tunnel which follows from a well-established and city-wide public art programme of which it is an extension. In the case of the Detroit People Mover, a new (1980s) system which serves the central business district, art was used to extend the ambience (and implied safety) of corporate space to its stations. But a number of questions arise from all this: to which publics is art in public transport addressed? and is it possible to use its location in the public realm as a space in which to address social criticism?

Art in public transport tends to assume a 'general public'. In this it shares one of the difficulties of conventional public art: in appealing to a non-specific public it reflects the ideas and tastes of its advocates. But there are two ways in which art in public transport addresses particular publics: the appeal to corporate customers, and the alternative appeal to specific communities, through, respectively, design and art which reflects corporate culture, and community-related arts.

Corporate art on the railway

Alexander Beleschenko's tall three-sided *Glass Column* in Apex Plaza, a corporate development adjacent to Reading mainline station (blurring the edge between public and privatised space) is a demonstration of new technologies in glass, using a mix of decorative techniques and lenses to create a kinetic effect so that, in the words of a publicity brochure 'for regular users of the station, the appearance of the Column is always slightly different, while for the first time visitor to Reading, the sheer scale and drama of the piece must surely impress' (BR, n.d.); an alternative interpretation might see the glass as affirming the opulence of corporate space, or as promoting the railway as a high-tech company through an association with leading-edge glass technology. What seems evident is that, whilst British Rail's art programme had (as with Atherton's *Platforms Piece* at Brixton, see Figures 25 and 26) been, until the late 1980s, aimed at a sense of community ownership, *Glass Column* marks a shift to an appeal to its executive market.

An appeal to the same market is made by Jean-Luc Vilmouth's *Channel Fish*, at the Eurostar terminal, Waterloo International, designed by Nicholas Grimshaw. The fish 'wiggle' when trains arrive and depart, using computerised technology to control their eel-like forms. The budget for the fish was around £200,000, although their commissioning was an afterthought in this award-winning[29] example of architecture which both denotes itself and expresses the mystique of high-speed travel in its clean lines and successions of space; although the structure is essentially engineering, in the tradition of Brunel's 'train sheds', its art, rather than that of the art added to it, is in using light and reflection to conjure spaces which have the resonance of a cubist painting. But the kind of aesthetic response stated in the last few lines is part of bourgeois culture, and implies a particular public, mainly the executives for whose travel the direct London–Paris link was built. Outside the concourse, homeless people beg for money as they have for several years.

Community art in transit

In other cases of commissioning for mainline stations, such as Brixton, or in some projects in the New York Subway, the public for art is defined more in terms of a local community; hence the work of Lee Brozgold with students of Public School 41 at Christopher Street, and Faith Ringgold's *Flying Home: Harlem Heroes Downtown* for 125th Street,[30] for which there are no equivalents on London Underground.[31]

For the Boston T, winning public confidence was particularly important: following the demolition of low-income housing to make space for a freeway in the 1970s, most transport development, including the relocation of the Orange Line underground, was met with hostility, whilst art was seen as a non-contentious way to regain confidence: 'Public art became a tool to help these residents cope with the change' (FTA, 1996b: 20). The organisation UrbanArts worked with 800 local people of all ages to gather oral history and create visual documentation of environmental change, as well as producing conventional (topographic) art, and poetry inscribed on granite slabs in stations; the FTA claim:

> The Orange Line projects suggest that public art can do more than enhance public space; it can advance a more far-reaching role in the social and economic revitalization of urban neighbourhoods. They demonstrate how communities can use public art to deal with conflict creatively and constructively.
>
> (FTA, 1996b: 21)

This sounds progressive, but art is brought into the service of a dominant planning approach, so that communities are allowed to 'cope' with change but not determine it. This is not to argue against community-based arts, and UrbanArts work within a tendency towards art as social process, but it is to question who writes the agenda for the change which art helps manage.

A reappraisal?

In Seattle, a city known for affluence rather than social problems, the integration of art in the construction of the Metro Bus Tunnel in the downtown area (in which it is a free service), from 1984, followed from an extensive programme of commissioning through the city's adoption of a Percent for Art policy eleven years previously.[32] A 1 per cent allocation of construction costs in the Metro Bus Tunnel realised $1.5 million for art, through which lead artists were employed to collaborate with engineers and architects during the design stage of the scheme, their brief covering the siting of art, colour and lighting. Jack Mackie was one of five lead artists, working on the Westlake Station and introducing the work of

Fay Jones (mural), Bill Whipple (clock), Gene Gentry McMahon (mural), Vicki Scuri (ceramic wall tiles) and Roger Shimomoura (mural). Kate Ericson was the lead artist at Pioneer Square Station, making a clock from found objects and introducing work by Mel Ziegler (clock), Laura Sindell (ceramic mural) and Garth Edwards (cut-out metal figures incorporated in railings).[33] More recently, artists have been included in the design team for new buses, though not without difficulties in finding a common agenda with employee representatives (FTA, 1996b: 26–7).[34] Artists, including Leila Daw, were also employed in the design of the light rail system for St Louis, where passenger loads have exceeded expectations (FTA, 1996b: 22); the intention was to adapt standard functional elements, such as bridge piers, into more aesthetically pleasing shapes – curving Ys rather than square Ts – and create a feeling of dynamic movement, though not all the ideas approved by the collaborative team were incorporated in the final specification.[35] It might be asked what artists can, given their training and aesthetic framework, contribute to the *design* of transport facilities – why should an artist make a bus design better than that produced by a bus designer? Brunel produced designs for bridges which have become regarded as classics of railway architecture, though they are simply elegant engineering solutions, and no artists collaborated with him. One response might be to see the artist as an 'incidental person' between professional designers and non-professional users. Reiter sees this as a problem:

> public art promoters have employed an argument that is inherently contradictory . . . the artist, as a non-professional, is cast as being in tune with the needs and concerns of real people . . . [whilst] artists enjoy a sort of hyper-professional status by virtue of the authority that comes with the concept of authorship.
>
> (Reiter, 1995: 62)

Yet if the artist's role is not as a kind of super-visually aware user, it becomes that of the 'expert', along with all the other professionals who extend but seldom question the structures of power which determine the city.

Is there another possibility? There are a small number of precedents, from the use of feminist iconography by Siri Derkert af Ostermalmstorg station on the Stockholm Metro (Greene, 1987: 56) to Tom Otterness' sculptures for 14th Street Subway station in New York, to be installed in 1997. Otterness, like Atherton at Brixton, uses bronze, and creates images with an edge of social comment as well as humour; an alligator (referencing the urban myth of alligators in the sewers and subways) emerges from a manhole cover to swallow a figure whose head is a bag of money, two figures attempt to saw through a structural support (Figure 42), a man finds a penny in the street whilst two uniformed figures sweep up heaps of coins.

There is also another side to metro art which uses the past. Jean-Paul Laenen's 1978 photomural *Métrorama* at Aumale Station in Brussels records the Anderlicht

FIGURE 42 Tom Otterness, figures sawing a column, for 14th Street Station, New York

neighbourhood 'redeveloped' when the subway was built, and in Jan Sierhuis' *Demolition Ball* at Amsterdam's Nieuwmarkt Station, an iron demolition ball is part of the piece, poised against a fragment of brick wall, over a photographic image of figures on the demolition site, referencing the development of which the station is part. Sierhuis states: 'The hanging up of the demolition ball . . . is a lasting token of protest' (Greene, 1987: 58). These critical works, whilst aesthetic and integral to station design, suggest possibilities for art which are not complicit with the agendas of the dominant institutions of society. Similarly, the posters, changed to reflect issues debated in the National Assembly, by Jean-Charles Blais on the Paris Métro, intervene in a critical territory. Perhaps, in time, there will be possibilities for art using the technologies of inter-active video and the internet to emphasise the location of stations in the public realm. Meanwhile, protests against new roads gain strength and public transport is seen as increasingly vital to urban sustainability, at least until a more radical approach based in mixed-use regeneration lessens the need to commute. One of the spontaneous events surrounding Rachel Whiteread's *House* was the fixing to it of posters against the M11 extension in Wanstead (Figure 43).

FIGURE 43 A poster against the M11 on *House*

7

ART IN HEALTH SERVICES

•

INTRODUCTION

This chapter concerns art in health buildings, which, like art in transport, has become a specialism within the field of public art; whilst universal health care and public transport are social 'goods' (though health care tends to be a more volatile issue, and in both fields arguments are advanced that artists and crafts-people should be involved in the design as well as decoration of facilities), there is an obvious difference in that people use hospitals only from necessity, and often in fear of illness and its consequences. Perhaps there is also a difference in that scientific medicine implies that its effects can be measured, suggesting that the contribution of art to healing might, too, be systematically evaluated. At the same time, the hospital is a recent institution with its own socially produced, influential and contestable culture; as the hospital has increasingly contained and organised, or medicalised, the events of birth, illness and death, its culture has been questioned through new approaches to care and a new interest in mind-body medicine.[1] In the UK, reorganisation of the National Health Service (NHS) and realisation that large hospitals are less efficient as well as more daunting than local provision in human-scale environments, has meant that some aspects of health care are returning to community settings.

This chapter is not a general history of art in institutions, being specific to art in the modern acute and long-stay hospital; the acute hospital combines the functions of the infirmary and the clinic, and the long-stay hospital continues those of the 'asylum'.[2] The chapter considers cases of visual art in hospital design, asks whether its contribution to patient care can be evaluated, and suggests that art may be either a cultural addition to hospital environments, or part of a change in the culture of health care. A case of art in a community health setting will be discussed in the following chapter.

THE ROLE OF ART IN HEALTH CARE

The fear of illness, pain and death remains a pervasive emotion in western society and the case for art in hospitals begins from the ways in which this fear is addressed; if modern medicine is a defence against death, then art in hospitals is a defence against the institutional environment in which medicine is delivered. As such, it may either be supportive of the institutionalisation of illness by seeking to lend its environment a greater acceptability, or, it may challenge the culture of modern medicine.

Art that prepared the soul for death as a natural threshold to the next life was a presence in religious foundations for the sick from the Middle Ages, but art in hospitals begins to take a modern form in Hogarth's murals, *Christ at the Pool of Bethesda*, of 1735, and *The Good Samaritan* of 1737, in St Bartholomew's Hospital, London; few patients see Hogarth's pictures, on the stairs to the Great Hall and a room now used as the Consultants' dining room, which were produced for a professional audience familiar with the narrative, the pool of Bethesda being a place of healing. The figures gathered around the pool are used to depict the symptoms of disease categorised by the emerging secular science of medicine[3] with its ability to diagnose and treat illness according to the taxonomy of symptoms, and its new approach to death, which is no longer carried within each person,[4] but an 'unnatural' force against which medicine offers protection.[5] Since the eighteenth century, according to Ivan Illich, 'a gigantic defense program waging war on behalf of "humanity" against death-dealing agencies' (Illich, 1976: 202) has been established, in which technical prowess has replaced the sanctity of life, and intervened in the acceptance and realisation of mortality. Illich continues: 'Through the medicalization of death, health care has become a monolithic world religion whose tenets are taught in compulsory schools and whose ethical rules are applied to a bureaucratic restructuring of the environment' (ibid.: 205–6), and a recent research paper states:

> When a person is admitted to a hospital, he or she must almost completely abandon those roles characteristic of non-hospital life and assume the status of someone who is 'sick' or 'injured' . . . circumstances [which] involve a loss of personal control, along with a concomitant ambiguity regarding the individual's personal responsibility for their progress through treatment.
>
> (Winkel and Holohan, 1985)

This process, in which the control of death or recovery becomes professional rather than intimate or familial, has produced the typically impersonal environment of the hospital; a parallel development has medicalised depression and other behaviour which challenges social norms. The case for art in health buildings today tends, consequently, to be based in a desire to 'humanise' these buildings by adding a cultural dimension. But this imposes a double burden because the culture of institutions is expressed by more than the buildings they occupy, and hence art is required to both add an aesthetic dimension to functional architecture, and minister to the feelings repressed by modern medical practice by being expressive and individualist. Art also relates to the world outside the hospital, through the depiction of local scenes, local history, nursery rhymes or cartoon characters, or by being 'modern art'. This is a strategy of distraction assuming quite a lot about art, not least its general appeal, possibly replicating the taste of members of hospital art committees whilst leaving intact the institutional ethos and professional hierarchies which determine the character of the building thus humanised.

Art in the Renaissance orphanage, or the Victorian infirmary, derived its value from a wider culture in which matters of the spirit were addressed through art as well as literature, music and prayer. In the second half of the twentieth century, art in hospitals is a matter of architectural embellishment, sometimes, particularly in the 1970s, of attempts to impose a kind of instant cheerfulness – for example by painting clowns, parrots, characters from Disney and 'jungle scenes' using garish colours in childrens' areas.

The origins of recent hospital art (or arts) projects are partly in community-based art and partly in conventional public art,[6] each suggesting a difficulty: the artist seeking to relate to hospital communities needs considerable social skills to contend with the negative emotions felt by people in hospitals; and art which reflects the taste of experts, such as the hospital art curator or committee, will simply strengthen the hold of the institution over its users. The duality is still present, as art in acute hospitals reflects public art's concern with a physical site, and art in long-stay institutions tends to be a form of community arts; in the USA, the pattern is slightly different, with more emphasis on the performing arts whilst participatory art projects have been developed particularly in cancer care.[7]

Art in NHS hospitals has tended to adopt a position affirmative of institutional structures, and continues a tradition of architectural embellishment from the nineteenth century; many Victorian institutions, such as the General Infirmary at Leeds designed by Gilbert Scott (1862), were decorated with tiles,[8] stained glass and stone or terracotta ornament, and were cases of progressive architecture in their time: the large, open-plan 'Nightingale' wards where air could circulate were thought to reduce cross-infection, and ensured constant observation.[9] Despite their image of authoritarian drabness today, some of the large asylums,[10] such as St Nicholas in Newcastle, were also embellished.[11] Several hospitals, notably St Thomas' in London and the Royal Victoria Infirmary in Newcastle have collections of tile pictures illustrating nursery rhymes, made around the beginning of this century by firms such as Doulton;[12] many others have stained glass in their chapels, and the Chelsea and Westminster Hospital has a Veronese relocated from a previous site, as well as a large collection of contemporary art.

During the past twenty years there has been an expansion of art projects in NHS hospitals, in context of the wider spread of art to non-gallery locations and supported by government advice,[13] whilst in the USA projects have been established in many University Medical Centres such as Duke University,[14] the University of Washington, University of Iowa and the University of Michigan, in independent hospitals and in the National Institutes for Health, and hospitals in the Netherlands and Sweden have applied Percent for Art policies to purchase art collections or commission works for specific sites. Although most projects in the NHS began by, for example, siting framed original prints in corridors and waiting areas, a few well-resourced projects, as at Leicester Royal Infirmary, began in the late 1970s to undertake commissions of site-specific art, such as a courtyard sculpture by Peter Randall-Page, and to employ artists to visually document hospital life. Peter Coles, reviewing the position in 1983, when efforts to introduce art into NHS hospitals began to be recognised by government, set out three main categories of hospital art projects: works of art commissioned as a complement to architecture; art as a factor in the social environment of a hospital; and artist-initiated residencies (DHSS, 1983: 2–3), concluding that 'the addition of works of art or the presence of an artist could stimulate healthy feelings, prevent any preoccupation with illness and encourage a therapeutic atmosphere' (ibid.: 3). Richard Cork similarly asserts that the quality of a hospital's environment is important in relation to patients' recovery – 'If they are incarcerated in buildings which make them feel uneasy, or intolerably nervous, the odds are stacked far more heavily against their ability to get well' (Cork, 1989: 191). In the 1980s, again in keeping with wider developments in public art, arguments were advanced that artists should collaborate with architects in hospital design. This was not always successful; Susan Tebby's coloured lattices painted directly onto the walls of a new outpatient area at Hammersmith Hospital, for example, although a collaboration with the architect John Weekes, were more or less destroyed by change in space utilisation at a later date. The use of standard construction materials as aesthetic elements was perhaps a more effective strategy, as with Dick Ward's series of pictorial duct covers in formica at Wansbeck low energy hospital, Northumberland,[15] and the integration of stained glass by Shona McInnes in the revolving doors of Queen Margaret Hospital, Dunfermline.[16]

More typically, art was commissioned during the construction stage of a building, as in the second phase of Musgrove Park Hospital, Taunton, where projects included sculpture and willow-planting in courtyards and decorative glass in the main corridor, and in the new wing of Perth Royal Infirmary, in which a tapestry by Fiona Hutchinson is sited. Many commissions were not in new buildings as such, but in refurbishment schemes, such as the national demonstration projects in the NHS discussed below.

In the USA, the National Institutes for Health in Bethesda, Maryland, included the Director of its art project on the design team for the new Childrens Inn, leading to an innovative approach to children's facilities: whilst these are often brightly coloured, it was realised that sick children feel more benefit from softly coloured surroundings, and the design palette is based on pale earth colours; each room contains original art; there are commissioned pieces of sculpture and textiles; and views to landscape are maximised in the building design.

NHS design, too, has changed in the 1980s and 1990s, producing human-scale buildings such as the Lambeth Community Care Centre, the West Dorset Hospital and, in primary care, the Chiddenbrook Surgery in Devon and Blackthorn Medical Centre in Maidstone.[17] Hospitals will continue to become smaller, partly in response to medical technologies such as laparoscopy which reduce the number of beds required by substituting day- for in-patient treatment, and partly as low-tech services are relocated to community hospitals, reducing the need for art as camouflage and increasing its role as autonomous aesthetic object or denoter of locality. In the case of participatory arts projects, the effect may also be to offer an alternative to clinical indifference.

The principal case made for art in modern hospitals is, then: to improve the appearance of the buildings so that patients and visitors 'feel better', whilst design may additionally improve efficiency, comfort and safety, and arts activities may, like complementary therapies, be an alternative to the monolithic attitude of the institution. Much recent growth in art commissions in NHS acute hospitals has followed from a concern on the part of health managers for 'quality care' which involves linked improvements to the service and the estate;[18] a business case can be made for art in this context as: creating a feeling of welcome and safety as the patient or visitor arrives; assisting in way-finding by providing landmarks or imaginative signage;[19] contributing to local distinctiveness;[20] and empowering staff to make choices about their workplace (whilst there are few choices open to them on other matters).[21] This role has two aspects: patient well-being, and the promotion of the service (in the UK of the NHS, in the USA of competing facilities).

ART IN THE NHS NATIONAL DEMONSTRATION PROJECTS

In 1989, the NHS Executive[22] funded the first of three annual rounds of national demonstration projects in acute hospitals;[23] the results were published in well-illustrated booklets[24] circulated throughout the service, and, together with many independent, local initiatives, contributed to recognition of the new approach to health design. The brief for the demonstration projects included provision for art and craft work, its particular form being determined by local project teams advised by an arts professional.[25] These projects were centrally funded to encourage innovative design, in which art was an element, in order to make visible the policy of providing 'quality care'. Whilst some improvements, such as computerised appointment systems, were behind the scenes, art and design were seen as effective in changing public perceptions of the NHS. Art was thus, on one hand, seen as a beneficial aspect of design, and, on the other, a way, with design, to create a more favourable impression of the NHS at a time when its management was being reformed, private health care was being widely advertised, and cost constraints were creating deficiencies in provision which were becoming politically sensitive and frequently taken up in the national press.

Each project determined its own use of art and level of budget; at Queen Elizabeth Hospital, Gateshead, the in-house designers transformed a courtyard into a new glass-roofed atrium and tea-bar, at the centre of a refurbished outpatient department. The public areas were carpeted, half the seating was moveable, the desk was low, open and finished in light ash veneer, and uplighters replaced florescent lighting; a long glass screen, necessary to create a barrier between the waiting area and the entrance lobby with its automatic doors, was identified as the vehicle for art, and made in stained and cast glass, titled *Sea Piece*, by Mike Davis and Kate Watkinson at their workshop near Durham. The design was chosen, as in all the demonstration projects, by the staff team. A nurse has since commented: 'The glass makes it look really different. It's the first thing people see, and some of them are worried on their first visit about what they might be told, and they can look at it while they relax and have tea' (NHSE, 1993: 21); and a patient stated: 'I've been gazing round here in wonder, it's like something from the future. It's absolutely great' (NHSE, 1991: 13).

Other art commissions in demonstration projects included colour photography by Terry Wright at St George's Hospital, Tooting, and Hillingdon Hospital, a three-part tile picture by Christine Constant at Cumberland Infirmary, a series of multi-part prints by Penny Berry at Peterborough District Hospital, a tapestry by Annie Sherburne at Whipps Cross Hospital, London, and a weaving by Sue Brinkhurst at Southend Hospital.[26] The demonstration projects were by no means unique, and art projects following similar principles had already been established in the pilot low-energy hospitals at Wansbeck in Northumberland and Newport, Isle of Wight, but the national profile of the demonstration projects established art as an integral aspect of hospital design, encouraging managers to initiate such approaches in their own sites, as in the commissioning of two tile pictures of Brighton by Ann Clark for the refurbished outpatient department at the Royal Sussex County Hospital in 1992. The demonstration projects also introduced some design features which may have more impact on patient care than art, for example the introduction of low, open desks which aid communication,[27] but today it

FIGURE 44 The interior of the new Outpatient Dept at Queen Elizabeth Hospital, Gateshead, with *Sea Piece* by Mike Davis and Kate Watkinson (Photo: Antonia Reeve)

FIGURE 45   An uplighter, with *Sea Piece*, beyond

FIGURE 46
Christine Constant,
tile mural of
Carlisle at the
Cumberland
Infirmary

would be exceptional for a hospital to reject the idea of art as an aspect of its environment, although there is no scientific evaluation of its impact.

EVALUATION

Anecdotal evidence in the form of patient and staff feedback is the most common kind of evaluation of art in health care.[28] Because medicine is a science, observation of the impact of care is a standard practice – drugs undergo clinical trials to assess their effectiveness (although there is a tendency for clinical procedures to be justified by tradition[29]) from which it could be argued that the contribution of art in hospitals can also be subject to performance indicators.[30] But the case for evaluation is not straightforward, even when matters of aesthetic judgement are excluded: it may be argued that the arts as a social process cannot be subject to the mechanistic measures of science;[31] or that the role of the arts is so integrated with design, and perhaps quality care, that it cannot be evaluated in isolation. A study, for instance, conducted by environmental psychology students from Nottingham University – which demonstrated variations in stress and alertness amongst patients in shabby and refurbished waiting rooms[32] and confirmed the attention given by patients to colour schemes, lighting and art – did not isolate the role of art within such contexts.

Some of the research data on the impact of the environment (rather than art) is, however, significant. Roger Ulrich, whose 1984 paper is the most often cited in this field, used medical records over nine years for patients recovering from gall-bladder surgery in an L-shaped ward in a Pennsylvania hospital, in a controlled study of recovery rates; a view of trees, being the only variable in the situation, reduced recovery time by about three-quarters of a day;[33] strong and medium doses of analgesic were also reduced[34] (Ulrich, 1984). Another study, conducted by a consultant anaesthetist in the NHS based on units in Plymouth and Norwich, showed that patients in intensive care units without windows to the outside world suffered a 48 per cent incidence of hallucination and delusion compared with 23 per cent for those in a similar unit with windows (Keep et al., 1980); an earlier study gives similar findings (40 per cent and 18 per cent) for two matched intensive care units in the USA (Wilson, 1972). These examples, and other papers,[35] confirm that the physical environment influences healing. But it is a leap of faith from that to say that art has the same impact, and a later study by Ulrich using abstract paintings produced no positive results (Ulrich, 1991); a study of wallpaper patterns in a unit treating patients with Alzheimer's Disease also failed to produce any significant conclusion.[36] Following reference to Ulrich's research, many interior designers, particularly in the USA, introduced landscape-based art in hospital design, for example by using large-format transparencies on lightboxes in enclosed rooms where scanners and linear accelerators are used, though these have not been formally evaluated, and a controlled study would require either a before-and-after model or a control group deprived of the images. On balance, it does not appear possible to make a case for art in hospitals based on scientific evaluation; studies are too few, those which are definite in their findings relate to the environment not art, and those which relate mainly to art tend to be anecdotal, whilst art is increasingly an integral element in health facilities and cannot be monitored in isolation from aspects of design, service and staff morale. There is also an argument that art is not subject to scientific testing in the same way as clinical practice, the latter not being beyond irrationality either (Ross, 1994). It is likely that the nettle of evaluation will never be fully grasped, and that the culture of health care will move on, accepting a mix of business and moral cases for the arts in context of a more humane, less functionalist delivery of care. Perhaps the main benefits might include the empowerment of staff[37] within a hierarchic structure, and the beginning of a challenge to the institutional ambience of the hospital. But cultural shifts cannot be evaluated through performance indicators any more than the saving of souls, for which art was made before modern times.

If art in acute care facilities is increasingly integral in design and hence difficult to evaluate, a more direct impact is found in the participatory projects of the long-stay sector, where treatment may be less interventionist and recovery is not always an option. An evaluation of the environment in psychiatric wards

(Kings Fund, 1977) states: 'In spite of all the efforts to organise industrial and occupational therapy and a variety of social activities, there were many hours when patients needed something else to do'. That 'something else' tends often to be art or music. In mental health hospitals, in which a core of highly dependent patients remain despite a general move to care in the community, residents have individual care plans devised and monitored by staff; these generally include: regaining or retaining self-esteem and the ability to make independent choices, effective communication and controlled mobility. Some evaluation studies have been carried out, mainly by arts professionals, though they tend to rely on anecdotal evidence gathered from interviews; a report on the work of the artist in residence and musician at Aycliffe Hospital in 1989, however, was researched by an NHS Quality Assurance Officer who observed the progress of individual patients, and stated 'People have grown in character and developed new skills and confidence, surely this is enough to justify the monetary cost' (Patterson, 1989). A study at Strathmartine Hospital, Dundee, provided evidence from staff on the effects of an eighteen-month residency by a visual artist, Lucy Byatt, in 1991–2;[38] the results were favourable, some staff seeing the artist as a link between discrete specialisms. Asked whether art projects had affected the delivery of care, eleven out of twenty said a clear 'yes', and three a clear 'no'. Amongst identified benefits were: that the creative potentials of residents were more recognised, and their quality of life improved; that the physical environment had been improved; and that staff could use simple and inexpensive art materials in projects of their own. These findings suggest art adds a creative dimension to institutional environments; other research indicates it has a psychological impact on staff, which in turn produces benefits to patients.[39]

CHANGING THE CULTURE

Evaluation of art in health care remains problematic. Perhaps there is a more important question – can art contribute to a change in the ethos of health care, a move away from the combination of eighteenth-century rationalism and nineteenth-century technology, towards an approach in which the patient's sense of self is restored?

A hospital Consultant wrote in 1994 that his patients did not wish to be better informed about their conditions, leaving 'their brains and their dignity at the desk',[40] illustrating the culture of health care to which art has hitherto been added, and that in this case at least it has not changed it. In contrast, Leland Kaiser has written: 'The future hospital has a two fold mission: (1) to become a healing community and (2) to heal its community' (Kaiser, n.d.); part of the healing required is the restoration of the psychological responses to death which are repressed by medical training but may be available culturally,[41] another part the learning of communication skills and empowerment of patients to create a non-authoritarian relationship between doctors and patients, the two being inter-connected.

The distancing of the doctor as an authority-figure is a function of the medical gaze – which sees the body of the patient as a representation, and is regarded by Foucault as an act of violence: 'to look in order to know, to show in order to teach, is not this a tacit form of violence, all the more abusive for its silence, upon a sick body that demands to be comforted, not displayed? Can pain be a spectacle?' (Foucault, 1973: 84) – justified by the need to retain a space between staff and patients who may die. Death must be distanced, abstracted, made a statistic, because it threatens to unleash repressed feelings;[42] on one hand is an operation of power through knowledge, as the medical gaze transforms symptoms into signs in an ordered system,[43] on the other a brittle coherence resulting from the splitting-off of mortality – so that, as Illich writes 'The good death has irrev-ocably become that of the standard consumer of medical care' no longer 'an exquisite and constant reminder of the fragility and tenderness of life' (Illich, 1976: 198–9) or a last rite of passage.[44] Illich ascribes the modern form of clinical prevention of death to industrialism;[45] Susan Sontag writes of TB and cancer in terms of an economic model,[46] and Foucault suggests a contractual relation whereby society (through its wealth) supports the infirmary for the poor but requires a dividend in the form of the research carried out on their bodies in the clinic (Foucault, 1973: 85), a model more or less perpetuated in the modern teaching hospital. And perhaps the legacy of the medical gaze and the denial of mortality also persist, the latter assisted by medical technologies which can prolong life almost indefinitely.

The abjection of the patient is challenged by recent initiatives for patient-focused care and patient-centred care, and by the introduction of a range of comple-mentary medicines and therapies in hospitals and primary care.[47] Patient-focused care[48] is a re-engineering of processes, for example bringing portable X-ray services to the patient instead of taking the patient to the service, delivered in multi-specialism 'bed floors' by multi-skilled staff teams, each responsible for all aspects of the care of a group of patients.[49] Although it is more efficient, it need not change the culture of an institution. Patient-centred care, on the other hand, is a more radical step developed by the Planetree community health organisation in California in pilot schemes in San Francisco and San José in the 1980s, and now implemented in several hospitals, including the cardiac ward of Beth Israel Hospital, New York, in the Griffin Hospital, Connecticut,[50] and the Mid Columbia Medical Centre, Oregon. In the NHS, Poole Hospital and South Cleveland Hospital are developing patient-centred care for local circumstances, and investi-gating the implications for design and staff training.[51]

Patient-centred care

The Planetree philosophy involves patient empowerment through education to enable patients to take an active role in their care, and in part through the creation of a physical and social environment conducive to a sense of healing which is deeper than clinical recovery. Both patients and staff have input to notes, the patient's observations sometimes giving information which is clinically important, more often giving a sense of partnership in care. There are implications for design, in removing barriers to communication, and the arts, which become part of the hospital culture, within its ethos of nurture.

The forty-nine-bed Mid Columbia Medical Centre, The Dalles, Oregon, is a Planetree community hospital employing 350 people who work with 200 volunteers, many local arts professionals, including painters, weavers, musicians and storytellers. The building, dating from 1958, was refurbished when the hospital adopted the Planetree philosophy in 1991 with a new glass atrium housing a grand piano played by volunteers at lunch-times; its furniture is motel-style and there are large plants, a waterfall and a small sculpture. The environment is free of institutional notices or other 'hospital clutter', and circulation areas retain the style of the foyer, even lift interiors being sites for original art. An aspect of the adoption of a hotel style of environment is that it enables the patient to feel more in control of choices, as they are when consuming hotel services; another is the retention of privacy, so that staff knock and await a response before entering bedrooms – all patients have single rooms in clusters round a nurse base and every bedroom has a comfortable chair, a shelf and chest of drawers for personal items, pine headboard on the bed, a piece of original art by a local artist and individual curtains and bed covers – and a two-way cupboard between the room and the corridor enables drugs to be allocated without taking a trolley into the room; where possible, patients are self-medicating. Small social spaces are plentiful, such as lounges, a patient library, kitchens and family dining rooms; the dietician makes coffee and bakes fresh bread each morning. Patients have the choice to use the restaurant, have food sent to their room, use the kitchen themselves, or have a meal cooked by a friend or family member. Staff and patients in a Planetree hospital can take part in the arts programme and benefit from complementary therapies such as massage and aroma-therapy. Bill Noonan, arts co-ordinator at Mid Columbia Medical Centre, states: 'Art is important because it helps meet the needs of the whole person, not just a person's body. We believe that by merging the resources of the mind and spirit, as well as the body, our patients may heal faster and more completely' (Miles, 1991b: 23). The combination of patient empowerment with complementary therapies and the arts begins to restore the intimacy of dealing with illness and mortality.

Planetree's philosophy may be the model for a future culture of health care. But care is increasingly moving out of hospitals into community settings, and this suggests further possibilities; an example of the role of the arts as part of the social process in which doctors participate is the range of projects, including visual art made in local schools as health education, and hosting a writer in residence, centred on the Withymoor Village Surgery in Brierley Hill since 1988. Malcolm Rigler, a General Practitioner in the surgery says: 'the belief is that you cannot be healthy unless you are being creative, and creative individuals make a healthy society. Being creative includes ordinary things like using words. The art of medicine is to know when to use medical intervention, when to be comforting and when to be challenging' (NHSE, 1994a: 34); and a doctor in the Blackthorn Medical Centre in Kent, has stated: 'we all want to see the patient in terms of the whole person. Some of this we are being taught by patients, who are all experts on their health' (NHSE, 1994a: 16). Blackthorn offers art therapy, music therapy, eurhythmy and gardening alongside conventional medicine, and undertakes rehabilitation work with people recovering from long-term psychiatric illness. These examples suggest the beginning of a new culture in health care which is less inclined to the military metaphors or authoritarianism of the infirmary, the clinic and their extension in the modern hospital, no longer confined to institutional buildings, and in which art is simply an aspect of that culture, not a cosmetic cheerfulness which denies the dignity of facing mortality, nor a replication of middle-class taste through beautification. But if the abstraction of the medical gaze is undone, then other institutions, other representations, are also questioned, including those of art, and other issues of empowerment arise, including that of urban dwellers to become experts on their city.

8

ART AS A SOCIAL PROCESS

•

INTRODUCTION

Conventional public art, as commissioned through Percent for Art policies, tends to be defined by its relation as aesthetic object to a physical site; in contrast, the emerging practices of public art in the 1990s constitute interventions in a public realm which includes the processes as well as locations of sociation. Patricia Phillips wrote in 1988 that 'some of the most fruitful investigations of public life and art are occurring in the most private, sequestered site of all – the home' (Phillips, 1988: 96), drawing attention to the role of the media and new technologies in intersecting the spaces of public events and domesticity; this is more than a blurring of spatial boundaries, in that the public and domestic realms are gendered, and for women artists the transgression of such boundaries is itself a resistance to patriarchy.

Suzanne Lacy terms these emerging practices 'new genre public art' (Lacy, 1995), and argues that they are rooted in the happenings of the 1960s but also informed by more recent discourses of Marxism, feminism and ecology. One element is a refusal of art's commodity status and the development of strategies, beginning with the lack of conventional art objects, to prevent colonisation by the art market; another is a reclamation of the role of the avant-garde artist as revolutionary. The value of new genre public art is, then, in its ability to initiate a continuing process of social criticism, and to engage defined publics on issues from homelessness to the survival of the rain forests, domestic violence and AIDS, whilst its purpose is not to fill museums, even with dadaist anti-art,[1] but to resist the structures of power and money which have caused abjection, and in so doing create imaginative spaces in which to construct, or enable others to construct, diverse possible futures. New genre public art is process-based, frequently ephemeral, often related to local rather than global narratives, and politicised.[2] It represents the most articulate form of a wider disenchantment with the artworld conventions still embodied by most public art during the 1980s.

This chapter summarises the discourse out of which new agendas for and practices of public art are emerging, beginning with the writings of Suzi Gablik, in whose *The Reenchantment of Art* some of the (mainly women) artists who have become known as new genre public artists were first profiled, and Charlene Spretnak, who proposes, like Gablik and simultaneously but independently, a paradigm shift away from Cartesian dualism (seen as a basis of the splitting-off of art from life) towards a more participatory ethos; it then selectively considers projects in relation to social health, cultural diversity, and environmental aware-ness, to convey something of the breadth of a field in which there are many local initiatives.

ART AT THE END OF MODERNISM?

The legal case concerning the removal of *Tilted Arc* from Federal Plaza in 1989 demonstrated the bankruptcy of late modernist art in terms of social relatedness.[3] A critique of modernism as a masculinist genre which failed to deliver the promises of its optimism is found in varying forms and degrees in the writing of Suzi Gablik, Eleanor Heartney, Lucy Lippard, Arlene Raven and Suzanne Lacy. To briefly note two positions: Gablik argues, in *Has Modernism Failed?*, that (late modernist) art has no coherent priorities, persuasive models, or means to evaluate itself, concluding that for the postmodernist mind, all is hollow (Gablik, 1984: 17);[4] and Heartney that ecology and ecofeminism constitute positions from which the connection between western culture's exploitation of women and its exploitation of the earth, and its anthropomorphism and patriarchal bias, are evident (Heartney, 1995: 141–2).

But in the ruins of modernism and the failure of the modern project for an ever-better world, are two possibilities: the deconstructive, in which meaning and value are seen as aspects of an outmoded approach to a world in response to which only irony and cynicism are now possible, however much this might inhibit action; and the reconstructive, in which, accepting the need to relinquish the privileging of patriarchy which determined the framework for the modern (and Enlighten-ment) project, new approaches to meaning and value are still sought, however difficult this might be. Charlene Spretnak writes:

> deconstructive postmodernism draws attention to the ways in which overarching concepts are actually culturally constructed and are not the universal truisms that most people assume. In its extreme forms, decon-structive postmodernism declares that meaning itself is impossible
>
> (Spretnak, [1991] 1993: 12)

and Gablik, in similar terms:

> This pervasive need of the deconstructive mind to know what is not possible anymore would seem to represent an absolute terminus in the

'disenchanted' modern world view; the self-checkmating of a now dys-
functional but apparently immovable dominant social structure. Deconstructive postmodernism does not ward off the truth of this reality, but
tries to come to terms with its inevitability, in what are often ironic or
parodic modes that do not criticize.

(Gablik, 1991: 19)

For Gablik and Spretnak, the outcomes of the governing western way of thinking
are so threatening to survival that irony and parody are insufficient, an intellectual
posture not enough, and a new paradigm is required which enables social intervention, though this intervention may be first in the self (in order that confusion
is not replicated in action). Gablik writes, in a key passage in *The Reenchantment
of Art*:

In our present situation, the effectiveness of art needs to be judged by
how well it overturns the perception of the world we have been taught,
which has set our whole society on a course of biospheric destruction.
. . . I believe that what we will see in the next few years is a new paradigm
based on the notion of *participation*, in which art will begin to redefine
itself in terms of social relatedness and ecological healing

(Gablik, 1991: 27)

and Spretnak in *States of Grace*:

The acute suffering of the Earth community instills urgency in this
work. The conceptual liberation of the postmodern moment engenders
possibilities. The cosmological context grounds us in the sacred whole.

(Spretnak, [1991] 1993: 32)

Gablik and Spretnak, then, set out a 'reconstructive' alternative in which there is
the possibility for a reclamation of meaning and value in art and life and a recovery
of a sense of the sacred;[5] these are not self-contained or aesthetic positions, but a
foundation for engagement.

Gablik argues that the environmental crisis follows from, rather than simply
coincides with, modernity's value-free scientific knowledge, itself an extension of
Cartesian dualism, and that the goals of growth, power and domination are not
sustainable; Spretnak argues that the assumptions of the modern worldview have
caused 'ecocide, nuclear arms, the globalization of unqualified growth-economies,
and the plunder of the Third and Fourth World (indigenous) peoples' cultures
and homelands' (Spretnak, [1991] 1993: 4). For Gablik the crisis follows from
the splitting of mind from body which has produced value-free science and value-
free art. Against the model of splitting and separation, Gablik draws on a paradigm
of interdependence, drawn from emerging currents in physics and ecology (Gablik,

1991: 22), to propose a re-integration within the self and of the self in society. From this the individual becomes no longer an observer (as if) outside a disintegrating world, but a participating agent in, by election, its death or reintegration; art can act as a catalyst to such a process, even in small ways. Moira Roth, writing on Lacy's performances, for example, suggests that the visions of commonality and difference evoked in personal reception of the work remain after the event to inspire the participants with feminist visions and new hopes (Roth, 1989: 161). Gablik, for whom compliance is not an option,[6] cites the view of theologian David Ray Griffin that change is more likely when people are neither cynical nor utopian; and Lucy Lippard, writing in *Art in the Public Interest* on the alternative comic World War Three, notes that in even 'the direst tales, people make love, people give birth, people resist, people survive' (Lippard, 1989: 228), loosely paraphrasing Gramsci in opposing an optimism of everyday defiance to intellectual pessimism.

ART IN THE PUBLIC INTEREST

Amongst the critical responses to the aridity of modernity's end is the collection of essays in Arlene Raven's *Art in the Public Interest*, described by Michael Brenson as documenting a growing dissatisfaction among artists with the artworld's materialism and self-absorption (Jacob *et al.*, 1995: 17). Raven begins by asserting that 'Public art isn't a hero on a horse anymore', arguing that 'art in the public interest' extends the possibilities of public art to include a critique of the relations of art to the public domain (Raven, [1989] 1993: 1). The work of Jenny Holzer, Barbara Kruger, Alexis Smith, Hans Haacke, Krzysztof Wodiczko and others could be cited, along with Suzanne Lacy's performances, such as *Crystal Quilt* with elderly women in Minneapolis in 1987, Peggy Diggs' *Domestic Violence Milk Carton* of 1992, the work of The Art of Change and Platform in London, the *Culture in Action* project curated by Mary Jane Jacob in Chicago in 1993, the 'social sculpture' of followers of Joseph Beuys in Germany and the counter-monuments of Jochen Gerz; these initiatives, mostly managed by artists rather than agents, sometimes independent of arts funding structures and their sanitising effects, are not part of a common programme, but there are common strands implicit in their definition of public art as art which addresses public issues, and the use, mainly, of processes rather than objects to do so.

The Domestic Violence Milk Carton

The *Domestic Violence Milk Carton*, on which Patricia Phillips has written (Phillips, 1994a and 1995b), epitomises several aspects of new genre public art, particularly a transgression of the confines of public and domestic domains. The project, managed by Creative Time in New York, began with interviews with two women

FIGURE 47 'We Got It' – a candy bar designed by production workers, Sculpture in Action, Chicago (Photo: John McWilliams)

prisoners, both victims of violence in their homes, and with therapists, psychologists and healthcare workers. Diggs produced four designs for milk cartons, of which one with the text 'When You Argue at Home, Does it Always Get Out of Hand?' juxtaposed with a partially clenched hand and a telephone hotline number, was put into production by Tuscan Dairy in New Jersey; 1,500,000 cartons bearing the design were distributed through supermarkets in New Jersey and New York during January 1992. Phillips writes: 'While the methods of dissemination marked a significant moment in public art practice, the artist (or any of us) can only speculate on the effect of this far-reaching, short-lived project' (Phillips, 1994a: 16), concluding that despite the throw-away nature of the cartons, some moments of individual recognition occurred.

Diggs adopted the strategy of designing for mass production (against the uniqueness associated with high art) and questioned the notion of site by working for a 'space' between the supermarket and the home; for women artists this is particularly important, as a way out of the restriction of the domestic realm assigned them by patriarchy. Carole Vance writes, in context of art and AIDS, that the distinction between public and private realms is false because neither has a hard boundary. She adds that the most personal decisions are influenced by public laws and policies, and that 'In struggles for social change, both reformers and traditionalists know that changes in personal life are intimately linked to changes in public domains' (Vance, 1994: 99). Diggs takes violence in the family and turns individual into public narratives, offering a message completed by the personal interpretations and responses, including use of the hotline number, of those it is designed to contact.

SOCIAL HEALING

Diggs deals with a violence contained within the spaces of the nuclear family; another kind of violence, in acts of depersonalisation, takes place in society's penal, educational and medical institutions, collectively a kind of 'third space' beside the public and domestic domains; whilst the stated purpose of such institutions is to care and heal, the effect is also to produce a hierarchic structure and exclusionary professional ideology or, as Illich writes in *The Limits of Medicine*, to 'medicalise' tracts of life. In response to this elimination of an individual basis for care in favour of a 'war on behalf of "humanity" against death-dealing agencies' (Illich, 1976: 202), in which the patient becomes a ground for the representation of disease, there is a growing recognition that in many areas, care, and perhaps a kind of healing, can be accomplished through sociation, and that this may involve art.[7] At the same time, issues such as homelessness and AIDS cut across the categories of health and social value, addressed, for instance, through John Malpede's workshops with the homeless of Los Angeles (Burnham, 1989), Krzysztof Wodiczko's *Homeless Vehicle Project* (Freshman, 1992: 55–75) or Martha Rosler's *If You Lived Here* (Wallis, 1991), and the *NAMES Project Quilt* (Weinstein, 1989). As Lacy stated in 1992: 'visual art today is becoming ... open to (1) community building through art, (2) social representation, and (3) new artists' roles in shaping the public agenda' (Lacy, 1992: 2). But as well as the problems of abjection and life with AIDS, is the everyday, and less publicised, business of depression and mental illness.

The Art Studios

The opening, since 1986, of three Art Studios in the north-east of England is the result of a partnership of local and health authorities and the Artists Agency in Sunderland. These look like any artists' group studios in redundant buildings, filled with individual and open work spaces, and are managed in the same way, the artists being represented on the management committee; the only difference is that the projects originated as responses to the mental health needs of local people in the three areas – Sunderland, North Tyneside and South Tyneside.[8] The Art Studio in Sunderland provides 'an environment which enables individuals to express themselves in a way that they would otherwise find difficult, whilst gaining

FIGURE 48 Wood-
carving in The Art
Studio, Sunderland

other skills in the process' (The Art Studio, 1996), and the three Art Studios offer people with little or no art training the space, materials, and support to enable them to create art; these might sound quite general notions, assuming that art is beneficial, though they challenge the separateness of art from everyday life and are understood in the particular circumstances of these studios.

The project began when it was found that some people who had received in-patient treatment in a psychiatric ward returned after discharge, asking to continue to work in the studio set up as part of an artist's residency; being able to do so met some of the needs of their continuing care, such as mixing and communicating with other people, and building self-esteem, as well as offering an opportunity to gain skills, and at the very least prevented the kind of bored isolation in which depression could easily become again unmanageable. Although the caricature of the western artist might, following Romanticism and the model of Vincent Van Gogh, or representations in popular fiction, seem that of an outsider prone, in Freud's view, to neurosis, for the artists in the Art Studios the reality is something more creative and interactive, more conducive to being accepted in society than the stigma of institutionalised mental illness; through regular events such as exhibitions and drawing classes, the Art Studios offer association with others with a common interest, outside the structures of mental health care. Hence the aim to provide a space to make art, and an opportunity to be empowered through the management

FIGURE 49 Detail of exhibition space, with work by artists of The Art Studio, Sunderland, 1996

of the space, meets a mental health aim, for the rates of readmission to institutional care have dropped since the Art Studios were opened (NHSE, 1994: 9).

One aim of the Art Studio in Sunderland is 'To elevate the role of visual art within mental health day care, and raise the awareness of mental health' (Art Studio, 1996); it does this through a training programme for arts centres, special needs schools, psychiatric hospitals, museum education services and other organisations in which staff need to develop skills in working with people with special needs, and there are plans to set up international exchanges in Spain, Poland and Belgium, and to host an international symposium. The achievements of members of the Art Studios include finding employment, gaining places in further education and becoming professional artists.[9] The most recent Art Studio is in South Shields, opened in a disused Synagogue in 1994;[10] it includes a gallery where members show work, organises exhibitions at external venues, and has weekly 'women's days' to encourage take-up of its facilities by women, including those who have suffered from domestic violence by men.

Following the success of the Art Studios, the Artists Agency has researched the potential for cultural responses to the social needs of old people, people with mental health difficulties and members of ethnic minorities in the inner city West End of Newcastle; the aim is to 'develop a strategic plan which identifies a holistic remit for arts and health' (Artists Agency, 1996: 7.1) through projects run by local people and employing a visual artist. The identification of ethnic groups as (assumed) communities of disadvantage is problematic; its justification in the report, which urges that projects grow upwards from the health and social needs of communities, is in the association of inter-generational conflict and isolation with mental health problems in the black communities.[11] The function of art in such situations is not to appropriate the vernacular arts of ethnic minorities to white museum culture, but to use the processes of art as a framework of empowerment, a catalyst to forms of sociation, for which there are models in new genre public art. Gablik writes of the work of Tim Rollins in New York: 'In the context of a sustaining environment, within a network of social support and mutual respect, things can be learned, close working relationships are formed, shared goals develop . . . lives may even be saved' (Gablik, 1991: 109); and the *Tele-Vecindario* project coordinated by Inigo Manglano-Ovalle as part of *Culture in Action* demonstrates the possibilities of working through new media with members of street gangs, decoding the images of mass culture and making new images of self identification, in process making a space for sociation within an environment of everyday violence. Around 250 young people a year are shot in gang-related incidents in the West Town area in which the project took place; it succeeded at least in part in bringing together members of different communities in the neighbourhood, and much of the equipment bought for the project remains in the possession of the communities. Although Newcastle and Chicago cannot be directly compared, there may be a model in Manglano-Ovalle's work which, if adapted to local circumstances and developed by local people, has wider application, using a technology widely accepted, because it is like television, in a way that art, which is like the privileged culture of museums, is not.

Art and AIDS

It has become as much a commonplace to talk of the 'AIDS crisis' as of urban crises, though the term figures in the rhetoric of the reactionary right. Coinciding with the spread of the AIDS virus, there was a desire, after the end of the cold war and its easy assumption of an external threat to western society, to find within that society groups who could be marginalised as a new enemy – an exercise in purification which also included the targeting of the NEA.[12] But AIDS is a continuing reality, not a sudden emergency, as much art about AIDS indicates in its documentation of the *lives* of people who are HIV positive, and its articulation of a political campaign for adequate health care and encouragement of safe sex. Whilst Simon Watney writes 'Many of us are now experiencing what amounts to a tidal wave of death' (Watney, 1994: 56), he also argues for cultural responses which make visible the lack of support for people living with AIDS – 'They were not rounded up and put into cattle trucks and taken away to death camps, but their need for properly supportive HIV education has been equally systematically ignored in all but a handful of countries' (ibid.: 6).[13]

Richard Meyer, writing on the graphics of AIDS activism in *But is it Art?*, begins by citing Gran Fury's 1989 poster 'Kissing Doesn't Kill: Greed and Indifference Do',[14] which mimics the style of Benetton posters, contrasting this to the wider development of activist graphics from the previous decade in which 'craftsy' methods such as woodcut printing were used to distance art posters from advertising, and that of the early 1980s, such as Esther Hernandez' *Sun Mad Raisins*,[15] in which consumer imagery is directly referenced through satire. The 1989 image also departs from the more conventional style of political graphics in the *Silence = Death* poster which appeared in Manhattan in 1986, subsequently used as a T-shirt and badge design, which adopts the pink triangle of Nazi persecution, asking in small type 'Why is Reagan silent about AIDS?', and the street-style of Donald Moffat's *He Kills Me*, in which repeated images of Reagan are juxtaposed with a black and orange target. Meyer writes: 'Gran Fury relied on visual pleasure, rather than terror . . . the collective was challenging the mainstream representation of the AIDS crisis . . . as alien, pathetic, monstrous, and/or murderous' (Meyer, 1995: 54) and cites Gran Fury member Marlene McCarty as rejecting 'victim photography'. The strategy is one of refusing the abjection offered by the dominant structures of power, the manufacturers of the 'AIDS crisis', through imagery utilising the means (and sophistication) of mass culture, whilst equally refusing to be contained within the professional circles of art practice.

The 1989 poster was commissioned for *On the Road: Art Against AIDS*, a touring 'exhibition' utilising non-gallery spaces curated by Ann Philbin for The American Foundation for AIDS Research and opening in San Francisco; other artists participating included Barbara Kruger, Cindy Sherman, Keith Haring, Adrian Piper and Robert Mapplethorpe. Kruger's billboard text stated 'Fund Healthcare not Warfare', and Mapplethorpe's *Embrace* showed two men clinging. Not all the artists were gay or lesbian, and some, like Kruger, had no previous involvement in the issues, their introduction being a strategy to widen the public debate, countering the ghettoisation of AIDS (Gott, 1994: 188). An aspect of this widening the territory was Gran Fury's 1991 *Women Don't Get AIDS – They Just Die From It*, produced (with text in English and Spanish) as a poster for bus shelters in New York and Los Angeles in a project organised by Public Art Fund. Not all city authorities were sympathetic to such campaigns, and *Kissing Doesn't Kill* was banned by the transit authority in Chicago in 1990, and both it and Mapplethorpe's poster were denied sites in Washington, DC, the same year.[16]

Gran Fury was invited to show at the 1990 Venice Biennale, creating a site-specific work including an image of Pope John Paul II with a text beginning 'The Catholic Church has long taught men and women to loathe their bodies and to fear their sexual natures . . .' and ending '. . . Condoms and clean needles save lives as surely as the earth revolves around the sun, AIDS is caused by a virus and a virus has no morals', and an accompanying reworking of an earlier image-text work *Sexism Rears its Unprotected Head*, which included an image of an erect penis; the work, described as 'counter-representation' (Gott, 1994: 195), was not accepted by the organisers and was impounded by Italian customs, ensuring wide publicity in the national press.[17] Meyer records that Gran Fury began to feel fatigued by the time of Clinton's election as President in 1992, and that only small-scale activity has taken place since, such as a simple text-only poster 'Do you resent people with Aids/ Do you trust HIV negatives? Have you given up hope for a cure? When was the last time you cried?'.[18]

But if the length of time since the first deaths from AIDS and the spread of the virus have produced what Meyer calls a 'normalization', a 'resignation and numbness' (Meyer, 1995: 81), the *NAMES Project Quilt*, described by Jeff Weinstein in *Art in the Public Interest*, has continued to expand since its beginnings in San Francisco in 1988, each new panel representing a person's death. Weinstein sees the *NAMES Quilt* as both art and ritual, a means of sharing and finding a structure for the sense of loss brought about by AIDS: 'A quilt is a map of devotion to doing and using as well as to completion' and 'What the reading of names does for me is fill out the artwork into something temporal, multisensual, communicative: a ritual' (Weinstein, 1993: 44, 51). The *NAMES Quilt* adopts the strategy, necessarily because those who are bereaved by AIDS are not only artists, of participation, so that professional artists act, in workshops, as catalysts to the creative work of non-artists; the same strategy was used by photographer Nicholas Lowe and writer Michael McMillan as artists in residence in settings linked to HIV in the north-east of England, in a project initiated by the Artists Agency and described in *Living Proof*. Lowe writes: 'Since AIDS I have been driven by the need to bridge the gap between the rhetoric of theory and the lived experience it sets out to define' (Lowe and McMillan, 1992: 11).

Amongst the projects of *Culture in Action* is *Flood/Diluvio* by Haha, a group of four artists: using hydroponics, a system for growing plants in mineral-fed

30TH AVE A 34TH ST QUEENS

TOMPKINS AVE AT HILL ST STATEN ISLAND

FIGURE 50 Gran Fury, bus shelter installations, New York (Photo: Timothy Karr, Public Art Fund)

water which prevents soil bacteria being transmitted to the plants, Haha created, with volunteers in an indoor community location, a garden from which green vegetables were harvested for AIDS care facilities; the garden was seen as a metaphor for a person with AIDS – survival depending in both cases on the maintenance of a fragile ecosystem and cooperation, and meetings on AIDS took place in the garden site, where educational literature was also displayed. Jacob writes that art 'in the guise of hydroponics' became a means to educate and initiate dialogue about AIDS, safe sex, social responsibility and caregiving, and a nonthreatening way to invite people to talk about such issues, whilst being part of an alternative agriculture of small-scale production (Jacob *et al.*, 1995: 91).

CULTURAL DIVERSITY

The acceptance of difference (of gender, race, class, sexual orientation and age) contributes to the end of social fragmentation; but it is a process of resistance as well as celebration, and it too requires bridges from social theory to art practice. Jacob writes: 'Difference is a key concept in the breakdown of the mainstream power structure', adding that racism, sexism, classism, and ageism remain commonplace in western democratic societies (Jacob, 1995: 98); and Dolores Hayden that: 'public culture needs to acknowledge and respect diversity, while reaching beyond multiple and sometimes conflicting national, ethnic, gender, race, and class identities to encompass larger common themes, such as the migration experience' (Hayden, 1995: 9). bell hooks argues that 'We have to create a kind of critical culture where we can discuss the issue of blackness in ways that confront not only the legacy of subjugation but also radical traditions of resistance, as well as the newly invented self, the decolonized subject' (hooks, 1995: 93).

Difference, however, in white liberal society, still tends to mean the privileging of a dominant interest, supported by the stereotypical images of consumerism and political advertising; the consequences of these attitudes are exposed in art such as the 1988 *Welcome to America's Finest Tourist Plantation* poster by David Avalos, Louis Hock and Elizabeth Sisco, and their 1993 *Art Rebate/Arte Reembolso* project,[19] or Border Art Workshop's performances titled *Border Sutures* of 1990 (Lacy, 1995: 205–6), all in San Diego, and social attitudes to difference investigated by Adrian Piper in New York (Lacy, 1995: 266–7) through performance pieces which confront people with manifestations of 'otherness' and, more recently, installations such as her video piece *Out of the Corner* at the Whitney Museum, New York, in 1990.[20] Mel Chin, at a workshop in Aspen in 1994, produced a 'parking permit' asking recipients to tick their car colour, race, sexual orientation and income, listing only non-white, non-heterosexual categories, and incomes below $6,000 and above $100,000, the majority thereby experiencing classification as 'other'.

As difference has become a prominent cultural and social issue, interest in art by people of colour which deals with questions of identity has increased, although the representation of people of colour in art's institutions, including public art programmes, faces the same barriers as that of women, and there is still a tendency for critical attention to focus on white interpretations such as Lucy Lippard's *Mixed Blessings* (Lippard, 1990), rather than black voices themselves.

Art and difference

Art about difference takes the forms of art made by white people which takes difference as its subject-matter, and art made by people of colour, including native peoples in North America and Australia, which seeks, as a first step, to reclaim histories which have been obliterated by the dominant culture. The reclamation of such histories is seen as the beginning of a wider process of empowerment, in that it establishes a cultural identity through which members of these groups build self-recognition. Amongst projects initiated from within the dominant culture are those organised by The Power of Place in Los Angeles, documented by Dolores Hayden (Hayden, 1995) and hence discussed only briefly here, and the Art of Change in London; amongst work by people of colour are the public art projects of Hachivi Edgar Heap of Birds in relation to Native Americans, and of photographer Carrie Mae Weems in relation to negritude, of which bell hooks writes in *Art on my Mind* (hooks, 1995: 74–93).

The Power of Place, a multi-disciplinary project in Los Angeles, uses people's memories of place as a point of departure for projects which involve the conservation of historic districts such as Little Tokyo, streetscape design, and the construction of monuments to 'invisible' historical figures, such as black ex-slave and midwife Biddy Mason (Hayden, 1995: 169–87). Hayden cites Lefebvre: 'Space is permeated with social relations; it is not only supported by social relations but it is also producing and produced by social relations' (ibid.: 41) and Castells, from *The City and the Grassroots*: 'The new space of a world capitalist system . . . is a space of variable geometry, formed by locations hierarchically ordered in a continuously changing network of flows' (ibid.: 42), to argue for a kind of archaeology of city form in which the appropriated meanings of place are revealed and exhibited, as when a monument to Biddy Mason is set up on the space where her house once stood.

An example of the work of The Power of Place is the preservation of historic buildings in Little Tokyo, near the site of the Museum of Contemporary Art's first, temporary building on which Barbara Kruger exhibited a text work in 1989[21] using a red, white and blue configuration referencing the US flag, and, after some negotiation, stating 'Who is beyond the law? Who is bought and sold?' and other such questions. The Japanese-Americans who live in Little Tokyo have memories of place which include deportation to desert camps in 1942, for which in 1988 compensation payments were agreed by the US government, reviving memories of the episode, as well as city development (for which read demolition) programmes from the 1950s to the 1970s. In 1986 a section of First Street was declared a historic district, preserving it from further destruction; between 1988 and 1989, Susan Sztaray, a planning graduate, initiated a public art proposal for the district, consisting of a pavement design representing the history of small businesses in the street. Sheila de Bretteville (the designer of the Biddy Mason wall and a co-founder of The Power of Place) developed the proposal into a design, later collaborating with Sonya Ishii and Nobuho Nagasawa, two Japanese-American artists. But whilst this project fits many of the requirements of alternative public art, in collaborating with a defined community to embody memories of place, in dealing with local rather than global history, and in its low-key intervention in the physical environment arising out of a lengthy link to the social environment, it also raises the question: once a community is empowered to the extent of regaining its cultural identity, what do they do with that empowerment to become co-designers of their city, to occupy that place where the form of the city is determined?

The Art of Change

Loraine Leeson and Peter Dunn, founding artists of The Art of Change, state 'The most progressive Futures will surely be those that propagate the creative value of fragmented power – decentralized, democratized – and of a culture made rich and vital from the many strands and threads of difference that are currently excluded or marginalized' (Dunn and Leeson, 1993: 147); now a group of six artists, they have worked with people living in London's Docklands since 1981 (originally named the Docklands Community Poster Project), providing art and design services to community organisations and action groups, working through existing community structures, and utilising digital image technologies to make work and devise public art strategies for local authorities. The photo-murals of the 1980s, a changing set on six billboard sites, charted an alternative and local narrative, in opposition to the publicity (and utopian narrative) for the Canary Wharf development;[22] the project assisted local groups to gain confidence and recognition, and the cultural production of the campaign became increasingly important, to the extent that the Development Corporation was described by the media in terms such as 'controversial', rupturing the illusion of harmony of its publicity material. Whilst referencing labour history, the photo-murals were not nostalgic re-creations of past places, nor monuments to alternative heroines like those of The Power of Place. The artists set out clear definitions for terms such as 'community' – 'All communities are essentially communities of interest', stating that 'the word is often associated with nostalgic or highly romanticized images of "place". This, in our experience, is not very useful – indeed it is counterproductive' (Dunn and Leeson, 1993: 142) – and of the choices necessary in engagement: establishing to which communities there can be a belonging; linking with them; working outwards from there, having a sustaining role in sharing common meanings and a transformative role in offering new meanings of experience. The group develop 'local narratives' against the grand narratives of international modernism and multinational economic interests (Dunn and Leeson, 1993: 143), defining the specificity of the voices suppressed by mainstream culture.

The Art of Change uses digital imaging technologies for resistance, as well as working with culturally diverse communities. *Tricks of the Trade* in 1992 used the image of the Trickster figure Anansi, one of those that have emerged from the diaspora of the slave trade, to connect a web of narratives in a billboard poster, and as the first image of a CD-Rom in a teaching pack for schools; *Celebrating Difference*, 1994, was a project with a school in an inner-city area where racial tensions are high, dealing with issues of culture and identity, commonality and difference, producing a photo-mural; and *Awakenings*, 1995, in collaboration with the education department of the Tate Gallery, takes Stanley Spencer's *The Resurrection, Cookham* and substitutes the life experiences of fourteen students of local schools for the images of figures rising from the graves in the painting, producing a photo-mural displayed at the Tate.

ART BY NATIVE PEOPLES

In the Americas and Australasia, the cultures of residual, native populations are taken over by tourism, their identities falsified in popular culture and subsumed in simulations which enhance the power of the dominant culture (which usually has the technology and the money).

In Pioneer Square, Seattle, in 1991, three kinds of art referenced the history of the people who owned the land on which the city was built in the nineteenth century. One was a bronze bust of Chief Seattle (the name an anglicisation) by James Wehn (1909),[23] of which Eleanor Heartney writes 'this monument masks a number of ironies, primary among them being the fact that the city honors with its name an individual and a group that it has historically attempted to annihilate' (Heartney, 1992: 20); another, a 1940 replica 'totem pole', in a form of carving belonging to a different group of 'natives' – it replaces an original stolen by white traders from Alaska and erected in 1899;[24] the third, *Day/Night*, a pair of enamelled steel panels with text and graphic images, on one side in English and on the other in Lushootseed, a Native American language, by Hachivi Edgar Heap of Birds and sited as a temporary work (Figure 51). The artist writes:

> In the city of Seattle there are countless references to our indigenous people. They are displayed in forms that range from professional football helmets to towering totem poles which are not even of the Puget Sound region, to the name of the city itself . . . Through these mismatched references chosen by the white men, we do not find institutionalized evidence of the living indigenous people.
>
> (Shamash, 1992: 40)

Pioneer Square is a tourist location, one of the oldest districts of Seattle, old warehouses being turned into small malls, and a place where homeless people and substance abusers spend time, amongst them Native Americans. The texts state:

FAR AWAY BROTHERS AND SISTERS WE REMEMBER YOU

and

CHIEF SEATTLE NOW THE STREETS ARE OUR HOME

echoing Chief Sealth's thought: 'Day and night cannot dwell together. The red man has ever fled the approach of the white man, as the changing mist on the mountainside flees before the blazing sun' (cited in Shamash, 1992: 40).

An earlier work by Heap of Birds was *Building Minnesota*, a series of forty printed signs, like street name signs in red on white, sited alongside the Mississippi in Minneapolis (1988–9), each stating

HONOR
[a name]
DEATH BY HANGING

followed in small script by:

DEC. 26, 1862, MANKATO, MN – EXECUTION ORDER ISSUED BY
PRESIDENT OF THE UNITED STATES ABRAHAM LINCOLN

each name referencing one of the forty men, for example Ma-ka'ta I-na'-zin (one who stands on the Earth), condemned by Lincoln for their part in the Dakota resistance to the 1862 Homestead Act, which reappropriated their lands to white settlers. When the government supplies on which they by then depended were late, the Dakota were told to 'eat grass'.[25] The signs formed part of a circle, symbolising the broken hoop of Native American myth; the site has other resonances in the pollution of the river, once a symbol of purification, and the view of modern buildings standing for the industrial state of the Union, which eliminated the Dakota as a nation.

Similar questions of indigenous culture, but also of its appropriation in development, are raised in Australia, by the development, for example, contested between whites and Aborigines, of the Old Swan Brewery site in Perth, discussed by Jane M. Jacobs in *Edge of Empire*. The brewery building became redundant in about 1985, and plans were made to adapt it as a tourist centre with shops and food outlets; the site was, however, the home of the Waugul serpent, though he proved difficult to place according to the rigid and alien cartography of white culture. Jacobs reports that whole tracts of land were of significance to the Aborigines: 'anxious remappings of Waugul were attempts to reassert the familiar order of colonial spatial logic in the face of the disorderly emergence of a repressed geography of Aboriginal meaning' (Jacobs, 1996: 120) as the site was incorporated into tourism. The solution proposed by the Aborigines, who occupied the site from 1989 to 1990, was to return it to parkland and public ownership,[26] a proposal they advanced by planting seeds and trees; the response from state and development interests was to incorporate token signs of Aborigine history in the site, such as landscaping, which included a coloured brick pathway for Waugul, or the siting of a collection of Aborigine art recently purchased by the state. But as

FIGURE 51 Edgar Heap of Birds, installation in Pioneer Square, Seattle

Jacobs writes, 'Aboriginal aspirations to return the site to Nature precluded the expression of "reconciliation" through marketable signifiers of Aboriginality . . . which could be incorporated into a commodified museum system' (Jacobs, 1996: 124). The protest failed to prevent development of the site for consumerism, and Jacobs also writes of a proposed centre for Aborigine culture and Australian wildlife in Brisbane, over which consultation will involve Aborigines, but not offer them economic control – 'the spectacle triumphs' (Jacobs, 1996: 135) and with it the ideology of colonialism translated into the re-presentation of indigenous people and their culture as objects of curiosity to white audiences. At the same time, the dualism which is the basis for the white man's pillaging of the earth and enables the substitution of representation for experience is not found in the cultures of the native peoples of America or Australia, for whom the relation with the earth is intimate.

ECOLOGICAL HEALING

Ecological issues are rising on political and art agendas; they inform Mierle Laderman Ukeles' work with the sanitation department in New York, Dominique Mazeaud's ritual cleaning of the Rio Grande, a study group and visit to Belize organised by Mark Dion with the Chicago Urban Ecology Action Group as part of *Culture in Action*, the work of the artists' group Platform in London, and the work of Helen and Newton Harrison, on which Heartney writes in *But is it Art?*, and art which is itself a process of healing the earth, such as Mel Chin's *Revival Field*. Common to many such projects is the implicit question stated by Heartney in summarising the Harrisons' work: 'Who shapes the ecological discourse and why?' (Heartney, 1995: 164).

Traditional as well as activist practices, however, have a role in raising the awareness of ecological issues amongst local publics. Peter Randall-Page's sculpture *Still Life* (1987) was made for European Year of the Environment,[27] one of several pieces which monumentalise endangered species, following his 1985–6 shell carvings for the Weld Estate in Dorset (Figure 52);[28] and Gablik notes Andy Goldsworthy's approach to nature, 'premised on respect rather than on domination' and using materials, such as leaves, twigs and ice, found in the landscape. The ecological organisation Common Ground has worked with both artists. Selwood describes their work as setting 'the aspiration which public art commissions in the future should seek to achieve' (Selwood, 1992: 26), and some comments made by Randall-Page about his work set it in the context of Gablik's proposed new paradigm of relatedness. In an interview, he has stated the motivation for his work as 'the belief that one can relate to the world as a participant rather than exclusively as a voyeur. Human nature is part of the natural world despite our sense of separation from it' (Jackson, 1992: 5).

New genre art for ecological healing

FIGURE 52 Peter Randall-Page, shell form on a footpath in Dorset for Common Ground

Amongst the artists included by Lacy under the heading 'new genre public art', two have a primary ecological concern: Dominique Mazeaud and Mierle Ukeles, both also profiled by Gablik in *The Reenchantment of Art*. Mazeaud derives her form of practice in part from the influence of Joseph Beuys and social sculpture; in 1993 she spent three weeks travelling in North Carolina meeting environmentalists, collecting visual material and samples of local soil and water, making *The Road of Meeting*, a strategy she calls 'netweaving' (Lacy, 1995: 263). Perhaps her best-known work is the *Great Cleansing of the Rio Grande*, begun in Santa Fe in 1987; this takes the form of a ritual walking, alone or with others, of the (usually dry) course of the Santa Fe river, a tributary of the Rio Grande, on the seventh day of each month, collecting garbage 'like a devotee doing rosaries' (Carde, 1990). Mazeaud keeps a journal of the walks[29] and quite soon stopped collecting 'treasures' from the river, as too much part of the art world's commodification of everything; Gablik refers to her work as making as surprising an intervention as Duchamp's exhibition of a urinal in 1917, on the grounds that it works from a myth of compassion, rejecting the patriarchal competition of the art world, and through 'a redemptive act of healing' forcing those who learn of the work to review 'the power operating in our cognitive and institutional structures' (Gablik, 1991: 122).

The project might be thought ephemeral, certainly makes a small impression on the Rio Grande, but according to Mazeaud 'All rivers are connected . . . People function in the same way. One way to activate these currents is through ritual. Rituals are icons of connections, they are the art of our lives' (cited in Lacy, 1995: 263).

Ukeles addresses the issues of waste recycling which threaten to overwhelm urban society, and since 1970 has been unofficial, usually unfunded, artist in residence at the sanitation department of New York City. An early performance work, *Touch Sanitation*, in 1978, began from the feelings of garbage collectors that because they handled waste they were themselves treated as waste; Ukeles shook the hand of every collector, saying 'thank you', like Diggs crossing the conventional boundaries of public and private space as the gesture was received personally. Gablik writes that her rare ability to 'empathically knit herself into the community' of sanitation workers conveys the joy of creativity (Gablik, 1991: 73), contrasting this with Serra's impositional manner; and Phillips draws attention to the contradiction of artistic freedom, foregrounded in Ukeles *Manifesto for Maintenance Art* of 1969, when male artists' success depends on 'a whole roster of personal and professional assistants' including 'wives, lovers, technicians, and fabricators that fuel the fantasy of independence' (Phillips, 1995a: 172). In 1973, Ukeles produced a street performance, *Wash*, which involved interviewing pedestrians outside a gallery and washing the pavement on her hands and knees in a stereotypical image of domesticity, of which Phillips writes: 'The obsession with hygiene served not only to illuminate the dark side of maintenance as domination and control, but also raised questions about the nature of public space' (Phillips, 1995a: 177); twenty years later, having developed a project, *Flow City*, to mirror back to New Yorkers the handling of their garbage (and in a 1983 piece literally mirroring back the public in a mirror attached to the side of a garbage truck), Ukeles orchestrated *Re-Spect* in Givors, France, in which 80 per cent of the city's sanitation staff were released by the mayor to participate in a 'choreographic display' of sanitation vehicles; the work culminated in an illuminated mountain of broken cobalt blue glass seeming to float (barge-mounted and illuminated) above the Rhône. For Ukeles, the recycling of waste is the key issue to a new society, which leads into questions not just of recycling, but the difference between need and want and the waste production of consumerism.

In Boston, a group of artists including Joan Brigham and Mags Harries has worked since 1989 on temporary projects using the waste sites produced by the city's replacement of its main freeway by a tunnel. Each year one or two exhibitions are held on such an outdoor site. Some projects have had an ecological message, drawing attention to bodies of water, native grasses, residual bird and fish life, the wind and the light of the sky. This represents a strategy of action in the discarded spaces of development rather than an attempt to find employment as decorators of development, and of artists working outside conventional structures of arts venues and funding, taking responsibility for their own self-presentation as one public in the city.[30]

An attempt to reintroduce wilderness landscape into the city is Alan Sonfist's *Time Landscape,* the permanent setting aside of a plot in Manhattan (at the edge of SoHo) for planting native species of tree. In London, the group of artists called Platform undertook a project in 1992 called *Still Waters* to draw attention to the lost rivers of the city;[31] elements of the project included a campaign to dig up a buried river, ritual actions to bring a river back to memory, and a sustainable energy assessment for the one river in London (apart from the Thames) still flowing. For Platform, more than The Art of Change, the visual products of their campaigns are less important than the discussions leading into them and the activity they generate, so that art becomes a kind of 'direct democracy'.

If Ukeles, Mazeaud and the members of Platform make art around environmental and ecological issues, others make art which is an ecological process, for instance, Mel Chin and Viet Ngo. Their work is not an extension of the land art of the 1960s and 1970s, which seldom had a content outside that of late modernism – Smithson, for example, was, according to Sonfist 'not really mediating between ecology and industry, but was simply hiring himself out to decorate an area of landscape the mining company had exploited' (Sonfist, 1983: 93) – but more a fusion of art and science.

Mel Chin began *Revival Field* near Minneapolis in 1990; it is sited in an area of toxic waste,[32] to which visitors have access only with protective equipment, and it inscribes a circle quartered and set in a square; the 'field' is marked out by a chain-link fence and planted with hyperaccumulators[33] which extract cadmium and zinc from the contaminated soil. It is intended in any future expansion of such fields that the minerals extracted by harvesting and incinerating the crops will contribute to the costs of the project, whilst restoring land to usefulness. *Revival Field* received a grant from the NEA after some contention as to its status as art – the grant was passed by an expert panel, but vetoed by John Frohnmayer, the NEA's Director, at a time of increasing hostility to the NEA from some Congressmen; Frohnmayer writes: 'Mel Chin explained that he was really using the earth as his palette and time as a part of the process . . . the aesthetic wasn't invisible at all – it just developed according to nature's timetable' (Frohnmayer, 1993: 238). Chin describes the work as a sculpture involving the reduction process. Unlike conventional sculpture, here the material is not seen and the process is biochemistry and agriculture, the invisible aesthetic being measured in terms of soil regeneration.

Viet Ngo, trained in Minneapolis as a civil engineer as well as sculptor, describes his work, which is a viable business, as 'a fusion of engineering, architectural planning and art' and responds to the question 'is it public art?': 'That is my intention, but I do not like to use those words because they segregate me from the working people' (Ngo, n.d.: 1). The sustainable technology Ngo has developed is a simple application of the ability of duckweed (lemnaceae) to absorb waste products from water, rather as Chin's hyperaccumulators absorb soil toxins, in place of chemical or mechanical methods of treatment; the weeds stabilise the biological reactions in the water whilst removing pollutants, and use methanes and sulfides as food sources, and either grow within a mesh which aids their management and harvesting or cover the surface of a pond in a short time if undisturbed. The ponds can be shaped – as with the 'snake' of nine parallel channels in which phosphorus, nitrogen and algae (the results of agricultural pollution) are removed from polluted water before its release into Devils Lake, North Dakota. The project, completed after two years of consultation and construction in 1990, uses the serpentine configuration to compress four miles of canal into sixty acres; there is a visitor centre at one end, and the harvested duckweed is used as an organic fertiliser and animal feed.

CONCLUSION

The idea that modernity has reached a point of non-viability, mired in its own destructive outcomes, produces the condition of post-modernity, in which previous forms of value, such as the privileging of dominant (usually white, masculine, bourgeois) cultural patterns and the construction of over-arching conceptualisations of history such as Progress (of which the idea of modernity is a case), are rejected, in some cases along with the concept of value itself. Baudrillard, for instance, states: 'the human race is beginning to produce itself as waste-product' (Baudrillard, 1994: 78). This is allied to an extension of simulation in mass culture, as in the products of the global 'state' of Disney; Heartney writes, noting that the 'jungle ride' at Disney World is to 'most people' more 'real' than the Amazon rain forest, that such simulations 'embrace a future in which nature is reinvented on a daily basis to conform to the requirements of technology and commerce' (Heartney, 1995: 141).

A number of arguments converge when Heartney's writing, and that of Gablik, Spretnak and Massey,[34] are taken together: if the earth becomes a series of simulations more 'real' than its actual surface then it no longer matters what happens to that surface; the privileging of representation, reliant on the visual sense, is one aspect of a masculine culture derived from the dualism of Descartes' rejection of the senses and the body in favour of an abstract mind, which is the only location in which representation, or simulation, could be 'real'. The achievement of Disneyworld, and shopping malls such as the West Edmonton supermall,[35] is to make the fantasy more real (and captivating) than 'reality'. Meanwhile, species, some hardly known, become extinct, which is forever, and, as Gablik writes:

'the basic metaphor of human presence on the earth is the bulldozer' (Gablik, 1991: 77).

In this context, engagement means either resistance, through art which is social activism, or the building of new, perhaps very local, models of healing and ecology within the old order — models which may be small or ephemeral but at least do not contribute to the deafening noise. Non-engagement, as in art which makes the city an aesthetic object, by default, is complicity in further destruction. But engagement begins from many differing positions. Gablik and Spretnak argue for a paradigm shift, a return to a sense of the sacred; Ngo sees his work in pragmatic terms, and writes 'I am no Gaian romantic; these places are working facilities to do the routine yet awesome task of cleaning our wastes' (Ngo, n.d.: 3). In a way, that act of purification might be seen as an extension of the Enlightenment project for perfection; but in another, it is a response to the urgency of the threat of a world overwhelmed by its waste, a deluge produced as the 'invisible' aspect of global capitalism, alongside the abjection produced by urban development. Neither waste nor abjection are simulations. In the end, both may be answered by a redefinition of human societies' needs, yet societies' wants are fuelled by the hollowness of culture at the end of the twentieth century, a space into which art might, perhaps, insert some remembrance of the mutuality which is necessary for human survival.

9
CONVIVIAL CITIES

•

INTRODUCTION

This chapter considers the relation of art to urban sustainability. In the formation of strategies for sustainability, it is recognised that ecologies are inter-dependent systems which are damaged in their entirety by the destruction of any element, and that empowerment is a strategy for conviviality, that is, for a society in which people of diverse races and classes and both genders live together without the dominance of one public over another.

Empowerment is a 'buzz-word' open to over-use, but it is also a real experience implying more than a theoretical concept of democracy and extending to collaboration between professionals and dwellers; in some cases, cultural identities are constructed as a beginning of empowerment, in others participatory processes of planning enable urban dwellers to co-determine city form and the uses of city space. Artists, at the same time, are working outside the conventions of public art to question dominant concepts of the city and engage people in local narratives and personal politics, creating an ambience of social criticism.

The two fields – urban planning and design, and art – are beginning to construct a dynamic in which each contextualises and interrogates the other. But, after three decades of conventional public art which has been largely complicit in the social fragmentation consequent on urban development, it is necessary that art, like architecture, is critiqued from a viewpoint outside that of the artworld, and its agenda identified as that of urban futures, not aesthetic reductionism or art market success. It then becomes possible for artists, designers and craftspeople to contribute, through practices which are decorative or activist, to a new urbanism based on the values of what Gablik terms 'connectedness', and what has been called by those who seek an ecologically responsible and communitarian society, 'living lightly upon the earth' (McLaughlin and Davidson, 1985: 22).

URBAN FUTURES

The urban future is painted bleakly. William H. Whyte writes in *City*: 'Ride the freeways and you see the consequence of a weakening center . . . a mishmash of separate centers, without focus or coherence . . . it is hard to see how it can do anything but worsen' (Whyte, 1988: 331). By the year 2005, more than half the population of the world will live in cities, inhabiting dwellings ranging from luxury apartments to sheds built of waste material and cardboard boxes;[1] the resulting mega-cities of both northern and southern hemispheres may have no coherent form in the sense of a centre from which the city radiates, or a river which is its organising principle, and may be fragmented into zones of wealth and abjection, safety and violence, largely invisible to the webs of new technology through which global corporations communicate with each other.[2]

The city of modernity, extending from the planned city of the Enlightenment to the post-modern city of executive decision, has been constructed on the basis of the 'way of thinking about the world' which Gablik links to Cartesian dualism[3] and its isolation of an autonomous intellectual self from a world which thus becomes a value-free ground on which fantasies of a world can be inscribed; this realm of fantasy also allows the invisibility of people who do not share the dominant fantasy, or who, being displaced, pollute it. One of its primary spaces of self-denotation is modern art, including the steel sculptures which embellish urban developments. Those that are shiny reflect the utopian gleam of development; those that are rusty affirm the otherness of art. Both kinds lend acceptability to development which is anti-social in that it destroys the patterns of sociation through which a neighbourhood has been liveable, or at least thought of as 'home'.

The disciplines of the built environment – planning, architecture, design and art – are faced, now, with a choice between continuing to construct urban wastelands and further sites of abjection, or developing a new ecologically and socially responsible urbanism, a model of the city which celebrates its diversity and does not seek to impose a monolithic perfection. Within the wider questions of what constitutes a city and for whose well-being it is produced, are questions of public space, urban design, and the articulation of the public realm; if art is socially beneficial, it is in the extent to which it addresses the needs of urban dwellers – by liberating their imaginations, contributing to the design of public spaces, and initiating social criticism to articulate the public realm – that such benefits can be identified. The role of art in the realisation of an agenda for urban sustainability is, then, twofold: applied art as an integral aspect of urban design, opening the possibility of a renewed practice of decorative art which can be joyful as well as functional; and art as a social process, resisting oppression, intervening in the public interest, transgressing the boundaries of public and domestic spaces, sometimes by creating events, sometimes by taking direct action. The two approaches set up a creative tension and are complementary rather than exclusive, each informing the other. There is no need for them to merge, and a possible third element is the presence in cities of artists for whom perceptions of diversity and the layering of history and memory replace the institutional prescriptions of an order which is in effect control. That most models for an alternative society are located outside the city leaves to those who choose to live in it the task of creating equivalent models which embrace the diversity of the city, not its utopian cleanliness, as its joy. Amongst those who have decided to live in a city are many artists.

Whilst the Enlightenment city, planned as a unified statement of power, excluded the 'dirt' of those facets of everyday life which were not convenient – the vagrant and insane, then the sick – the sustainable city incorporates what Sennett terms 'disorder'. Chantal Mouffe argues that this model of non-controlled diversity applies also to political life, making the notion of consensus problematic. She sees the use of deconstructive theory as rupturing any illusion of an overarching narrative through which society can be unified, thus keeping 'the democratic contestation alive'.[4] Ernst Bloch, in *The Principle of Hope*, expresses a desire for a mutuality of reflective and practical work which might be seen as an aspect of such an encounter. He writes that the kind of knowledge required for 'decision' is 'not merely contemplative, but rather one which goes with process, which is actively and partisanly in league with the good which is working its way through' (Bloch, 1986: 198).

A new urban discourse

A new, multi-disciplinary urban discourse is emerging, though still mired in the assumptions of a bourgeois public domain, but opening questions which, the more unpacked, will lead to increasingly radical positions. The new dialogue amongst urban professionals has produced the term 'liveable' applied to urban spaces, for example in the name of the organisation Partners for Livable Places[5] founded in 1977; a decade later, Allan Jacobs and the late Donald Appleyard, both then teachers of urban planning at the University of California at Berkeley, stated 'livability' alongside 'identity and control; access to opportunity, imagination, and joy; authenticity and meaning; open communities and public life; self-reliance; and justice' as goals for future urban environments (Jacobs and Appleyard, 1987, in LeGates and Stout, 1996: 169). Their alternative manifesto[6] follows, but is not derived from, the critiques of Jane Jacobs and William H. Whyte, both proposing a reclamation of the street as a place of informal mixing at the heart of urban society.

The development of a concept of participatory planning by Paul Davidoff extends the possibilities of urbanism from design to social process; Davidoff introduced the idea of the planner as advocate[7] in the late 1960s, at the time artists were rejecting the commodity status of art through happenings or becoming community arts workers. Today, as action planning receives renewed interest, artists are again turning to activism and collaborations with communities. One implication of this for both art and planning is a redefinition of the public realm beyond physical sites which does not abolish the significance of public space in a democratic society. Alongside the spaces of public access television and the internet, the public square or urban park is where society is mirrored back to itself in its diversity.

The city offers the diversified sociation termed by architect Christopher Alexander as a 'semi-lattice'.[8] For the dwellers of medieval cities the city offered freedom from daily labour on the land and feudal ties, freedom to become a political class, though the freedom applied only to men and merchants, for whose ease public space was constructed and the inscription 'city air makes us free' placed over the gates of the cities of the Hanseatic League (Sennett, [1990] 1992: 135; Girardet, 1992: 118; Sorkin, 1992: xv). Sennett, commenting on diversity in today's cities, sees the free appropriation of spaces for unplanned uses,[9] and the development of neighbourhood identities, as part of the attraction of urban life, describing New York as grasping the imagination as a city of differences, its population gathered from all over the world, though its edges are protected by disengagement: 'difference from and indifference to others are a related, unhappy pair' (Sennett: 128–9).[10]

Alternatives

Perhaps the state of anomie painted by films such as *Strange Days*, in which the urban and social fabrics no longer correspond to any pattern beyond the most contingent, in which mere aporia would be a luxury, is taken from the dominant script for the West's urban future, or that which most appeals to audiences made hungry for the spectacular; but alternative strategies have been proposed – design for environmental and energy conservation,[11] and the empowerment of urban dwellers to participate in urban planning and design. If there can be art in the public interest, so perhaps, through such participatory approaches to urban planning and design, there can be development in it also.

Cases of community-centred planning include the redevelopment of Boston's Columbia Point as Harbor Point, where high-rise blocks have been replaced by low-rise, high-density housing giving greater territoriality as well as an increased sense of streetscape (Goody, 1993),[12] the Coin Street development in London, in which residents selected the developer and controlled the allocation of units to commercial interests,[13] and the redevelopment of Holly Street in East London, where, again, high-rise concrete blocks are being replaced with a kind of urban village, although this in itself will not solve social problems such as unemployment and crime.[14] These cases remain within progressive but conventional, perhaps bourgeois, models of planning with their emphases on the territory of home as a demarcation of a domesticity which complements the publicness of the street, whilst more radical alternatives, such as Arcosanti in Arizona, a mini-city for 5,000 people designed by Paolo Soleri (McLaughlin and Davidson, 1985: 251–5), denote alternative patterns of settlement;[15] other models, such as Findhorn, have abandoned the city for rural self-sufficiency, but their values of community living remain instructive for future urban development in terms of the pattern of co-operative sociation they construct.

Goals for urban life

Jacobs and Appleyard set out the requirements for 'a good urban environment', whilst independently in the same year the Urban Design Group in London published an 'agenda' for urban design. Jacobs and Appleyard define liveability as 'a place where everyone can live in relative comfort . . . This means a well-managed environment relatively devoid of nuisance, overcrowding, noise, danger, air pollution, dirt, trash, and other unwelcome intrusions' (Jacobs and Appleyard, 1987, in LeGates and Stout, 1996: 169), which is like the purified city of Protestant taste that Sennett sees as repressing diversity – dirt is, after all, only matter out of place. The Urban Design Group recognises that values may clash in defining good city form, and sets out a more flexible minimum definition of a successful city as:

> a mixture of uses: living, working, shopping, and playing . . . access to different activities, resources, information, places for all sectors of the population . . . permeable to all, regardless of age, ability or income . . . protection and security . . . opportunity for people to personalise their own surroundings . . . private spaces offering the opportunity for personal expression and public spaces robust enough to accommodate changes by their users.
>
> (UDG, 1987: 34–5)

Many of the UDG's points are made also by Jacobs and Appleyard, who state, for example, the importance of personal space and expression, combined with common interest and participation, and the needs for access to opportunity which imply choices in housing, work and cultural provision.[16] From this urban dwellers will, they argue, be more likely to take part in decision-making, and in neighbourhood projects. They also propose that cities be increasingly self-sustaining in energy, and that urban design should cater for the poor as well as the rich, but in referring to the question of meaning they use a commonplace – that people should be able to understand the city through a kind of transparency of routes. Kevin Lynch also defined the legibility of cities this way, setting out elements such as paths, edges, districts, nodes and landmarks as aids to a 'reading' of city form (Lynch, 1960); but paths and landmarks may, like pieces of public art, mean anything or nothing to different publics for whom questions of ownership are more important than legibility. Dolores Hayden writes that an 'aesthetic approach to urban design often holds meaning only for limited cultural elites and affluent neighbourhoods, and it is still a top-down process' (Hayden, 1995: 235), and planner Miffa Salter argues that it is the relation of client, user and professional which exists first and into which the urban designer intervenes.[17] That intervention may be perceived by any party as on the side of the others, or not, again, in part, depending on the relative sums of cultural capital held.

Two further principles are stated by Jacobs and Appleyard: a certain density of population which creates a metropolitan rather than suburban ambience, and a return to mixed-use zoning, which is a factor on which all progressive voices agree. It was elaborated as early as 1961 by Jane Jacobs in *The Death and Life of Great American Cities*, in which she argues that safety in urban streets follows from 'an intricate, almost unconscious, network of voluntary controls and standards' (Jacobs, 1961: 32) and requires use through most of the day. Single-use zoning restricts the use of streets to the times of day at which people enter or leave the zone, and this desertion leads to fear, and fear, if accompanied by social deprivation and the dereliction of buildings, to crime, which in turn further depopulates the street and turns wealthier inhabitants increasingly to a world bounded by the garage, the freeway and the underground parking lot beneath the corporate fortress or mall.

PUBLIC SPACE

Emphasis on the design of public space within urban development, as required by the zoning regulations of New York following the impact on them of Whyte's research (Whyte, 1988), supposes it has a particular function which justifies public expenditure on its provision, that it is more than the space left over in plans, or 'sloip'. It also implies recognition that cities have a social reality as well as a physical form, that in a democracy public space matters because it is where sociation takes place. Most societies maintain at least a semblance of allegiance to the idea of public space, as they do to art, adopting, for instance, pedestrianisation schemes for historic districts, though both art and space can be co-opted for other purposes, so that, like the plazas of the Broadgate Business Park, or Hay's Galleria by the south bank of the Thames, they are signifiers of the rise of corporate culture, or, as with the 'streets' of shopping malls, signifiers of consumerism. Sorkin observes homogeneity – 'a single citizenship of consumption'

– and control in the mall, noting 'its obsession with "security", with rising levels of manipulation and surveillance over its citizenry and with a proliferation of new modes of segregation' (Sorkin, 1992: xiii). Harvey interprets urban decline in terms of the capitalist division of space into marketable and tightly controlled portions;[18] and geographer John Rennie Short compares the colonisation of urban spaces for corporate development with imperialism:

> In the imperial past overseas colonization was underwritten by the British Army and Navy. Now it is the police who defend the urban colonizers. It is not that crime is any more prevalent in gentrified areas, although the contrast between the rich and the poor does provide greater opportunities. It is more a case of the new middle classes having the right language and necessary confidence to demand better policing.
>
> (Short, 1996: 168)

adding that riverside enclaves offer greater protection against the disaffected. Murray Bookchin, an urbanist who differentiates the idea of the city[19] from that of a sprawling urbanisation which engulfs rural and urban areas alike, also writes of a sense of besiegement: 'the city *and* the country are under siege today – a siege that threatens humanity's very place in the natural environment' (Bookchin, 1992: 3).

The social life of small urban spaces

Jacobs' plea for a new urbanism has produced empirical research into the uses of urban space (leading to mechanisms for attracting publics back into it), rather than theoretical work on urban concepts. William H. Whyte's *Street Life Project* began in Manhattan in 1971, and used time-lapse photography to observe the uses of street corners, plazas and urban parks. With a team of students, Whyte recorded how many people, of which gender, singly or in pairs or groups, stood, sat or held conversations in each space. From the results it became obvious that some factors encouraged the use of space, and others discouraged it. The obviousness takes the form of people sitting where ledges, low walls or steps provide a place to sit, perhaps with a suitable vista or in sunlight, whilst setting spikes in such ledges and walls discourages their appropriation: 'It takes real work to create a lousy place. Ledges have to be made high and bulky; railings put in; surfaces canted. Money can be saved by not doing such things' (Whyte, 1980: 29). But other factors are not obvious, for instance that pairs of individuals were observed holding conversations in the midst of a flow of people rather than moving to one side, explained as a matter of maximum choice to continue or break off the contact; or that many people, when asked why they found Paley Park peaceful, cited the water-wall, whilst when played tapes of it identified the noise as that of a subway train or truck (Whyte, 1980: 48). Amongst devices which encouraged

FIGURE 53
Somewhere between
graffiti and art, on
a boarded-up
doorway in Seattle

the use of public space were moveable seating (which people like to shuffle as a
sign of possession), sunlight, trees and flowers, flowing water, provision of refresh-
ments and street music. Whyte noted that the aesthetics of building design
and the shape of a plaza were of little importance,[20] although the provision of
some kind of public space, preferably with planting, was a desirable attribute of

FIGURE 54 Paley Park, New York

development which should be (and now is) enforced through zoning regulations; this has produced a multiplicity of 'public spaces', but some of it remains sterile, and some, particularly the public atria of corporate buildings, are inviting only to some publics.

Two of the most densely used public spaces, for their size, were Paley Park (Figure 54) and Greenacre Park (Figure 55), both 'vest pocket' parks in mid Manhattan;[21] both are visibly separate from but open to the street, have extensive planting and moveable seats and tables, and have small cafes. Both are still well maintained and used by office workers in break times and by older, middle-income residents of streets in the low fifties on the East Side. Both also have water features; one study of Greenacre Park states: 'The major element contributing to this sense of relaxation and retreat is the dramatic waterfall that dominates the site visually and aurally . . . a large proportion of the people in the park can be seen gazing directly at the falls' (Carr *et al.*, 1992: 99). Greenacre Park has three levels, one under cover, which attract differing uses, the main street-level being that most used for sociation and the lower terrace in front of the waterfall for intimate conversations or privacy. Neither park includes art, and a challenge to the advocates of art in urban spaces is to say what it could add to spaces which are already convivial.[22]

FIGURE 55
Greenacre Park,
New York

Whyte, writing before critical attention to the spread of surveillance culture, discourages overt security precautions which are unwelcoming to the casual user, preferring, like Jacobs, to see the presence of people as a source of both security and enjoyment: 'What attracts people most, it would appear, is other people' (Whyte, 1980: 19), but his method is empirical, and his text tends to engage with questions

FIGURE 56 Zoning regulations ensure trees but not sociation – spaces remain too regulated and sterile

of utility rather than meaning. Whyte is also prone to a masculine perspective, or 'front-yard' view of urban form (Mozingo, 1989), in which the observation of other people, a kind of urban gaze, is privileged over conversation or revery.

Although Whyte's approach is person-centred, in that it seeks to identify the elements which encourage people to use public space, and Whyte does not see a threat in the presence of so-called 'undesirables', it remains within a physical definition of site. Hayden advocates a more radical attitude: 'we struggled to focus on social and political issues, rather than physical ones, as the centre of urban landscape history' seeking a shared authority through 'a willingness to listen and learn from members of the public of all ages, ethnic backgrounds, and economic circumstances' (Hayden, 1995: 235). This activism, parallel with the methods of activist art, sees the dweller as a participant in a process, whilst for Whyte the dweller is more often the object of the time-lapse camera's gaze. Hayden continues: 'This approach gives primary importance to the political and social narratives of the neighbourhood, and to the everyday lives of working people. It assumes that every inhabitant is an active participant in the making of the city, not just one hero-designer' (Hayden, 1995: 235–6); and Miffa Salter cites Lefebvre in arguing that it is on top of the city constructed by power and money that 'users' overlay social practices which may be more or less undetectable to some professional observers but are part of a delicate web of repossession.[23]

The work of Whyte's Street Life Project has been extended beyond his empirical base to include questions of community participation, and applied in many
cities in the USA[24] and recently the Czech Republic, by Projects for Public Spaces
(PPS), a multi-disciplinary group of professionals and researchers based in New
York. PPS works on the basis that involved communities are more likely to take
part in initiatives to improve their neighbourhoods and commercial districts, and
uses such methods as focus groups of fifteen to twenty people with a facilitator
and recorder to articulate the needs of defined publics. They have advocated art
within urban design, often to make change visible. Their definition of 'place' is
set amidst a matrix of sociability, activity, comfort and access, rather than
remaining a purely physical notion of site; they also emphasise the need for multi-
disciplinarity, since professionals concerned with particular elements of urban
development or regeneration, such as traffic management, architecture, landscape
design, economic development and planning, tend to work within disciplinary
boundaries which exclude each other as much as users, and offer in discussion
only the opportunity to indicate failure. PPS seeks to discover first the views of
users, who may lack the vocabulary to speak directly with planners, from which
to present a professional agenda which may include elements such as maintenance
and security, entertainment, street design and street furniture, art, traffic exclusion
and transport access and local markets.[25] Although PPS does not have the explicit
concern for marginalised groups of The Power of Place, normally being employed
by a municipal client, their approach suggests a basis for development in which
design is linked to sociation.

Empowerment of urban dwellers

The design of public space is, then, one aspect of a solution for liveable cities.
Another is the development of social processes, beginning with a kind of empowerment which enables non-professionals to work on an equal basis with members
of the professions of the environment and city authorities, in making underlying
decisions about development.[26] It is, in a way, an overthrow of the dominance of
institutions in public life and of the subservience of people to the 'tools' of society.[27]
Part of this process, which is inevitably political, involves decisions to create public
spaces, but it extends into the whole range of planning decisions and urban zoning.
Planner Peter Hall writes that between 1960 and 1970, planning became 'an
apparently scientific activity in which vast amounts of precise information were
garnered and processed in such a way that the planner could devise very sensitive
systems of guidance and control' (Hall, 1988, in LeGates and Stout, 1996: 386).
One of the problems thus created is that the supposedly neutral observation of
urban life became in time a form of imposition, just as Burgess' concentric ring
model became prescriptive rather than descriptive. Hall charts the challenge to

systems planning from both the political right and left, noting the work of Marxists including Lefebvre, Harvey and Castells; he summarises:

> the structure of the capitalist city, itself, including its land-use and activity patterns, is the result of capital in pursuit of profit. Because capitalism is doomed to recurrent crises, which deepen in the stage of late capitalism, capital calls upon the state, as its agent, to assist it by remedying disorganization in commodity production
>
> (Hall, 1988, in LeGates and Stout, 1996: 391)

which seems close to Gramsci's formulation of hegemony, in which the interests of the bourgeoisie are advanced through the institutions of the state (in this case its planning bureaucracy, provision of fiscal incentives for developers, and public art organisations). It also reflects Illich's prediction that capitalism will enter a period of crisis because its principles are inherently destructive: 'Almost everyone in rich societies is a destructive consumer. Almost everyone is, in some way, engaged in aggression against the milieu' (Illich, 1973b: 102). For Illich the solution is in the demythologisation of science, a reclamation of language,[28] and a recovery for everyday life of politics and law.

Illich is concerned to re-appropriate the 'tools' of conviviality. There is a further layer to the problem in that the methodology of planning privileges the representations of space (in Lefebvre's terms) of the expert, just as the law, according to Illich, privileges the dominant interests in society; yet, if a doctor describes his patients as 'experts on their health' (NHSE, 1994: 16), perhaps dwellers are also experts on their city and if so, their expertise begins in their awareness of the spaces around their bodies and the lattices of memory and appropriation they assemble as a personal reading of the city. From this it follows that the role of the planner becomes that of enabler, assisting members of communities in acquiring the vocabulary and information, added to the empowerment of community identity, to affect planning outcomes. There is a parallel between planning which involves community participation and art which engages with defined publics in participatory work, such as Mierle Ukeles' unfunded residency at the New York sanitation department, the performances of Suzanne Lacy, and the work of The Art of Change.

The Urban Design Group's agenda states: 'those groups who have been under-represented in the past, must be helped to express their views and achieve environments to which they can relate' and that urban design 'operates . . . to achieve community objectives through understanding and using political and financial processes' (UDG, 1987); these objectives relate to the notion of 'advocacy planning' proposed by Davidoff, which is increasingly adopted as urban crises are seen to be impervious to conventional planning methods. Davidoff writes that the advocate planner 'would devote much attention to assisting the client organization to clarify its ideas and to give expression to them' which may involve 'expanding the size and scope of his client organization', though the advocate's first role would be 'to carry out the planning process for the organization and to argue persuasively in favour of its planning proposals' (Davidoff, 1965, in LeGates and Stout, 1996: 426). Davidoff's model, based on courtroom interaction, has a number of problems, such as the initial need for funds to employ an advocate planner which may not be available in the cases of marginal communities or small, local organisations, and the nervousness of hitherto detached professionals faced with community engagement; it also depends on a pluralism in which the contesting advocates representing special interest groups arrive at solutions which are not mutually destructive. But the problems have not prevented some urban planners, and architects working in planning, from developing participatory methods; one example is the use of urban design action teams, or UDATs,[29] as a means to bring together dwellers, developers, officers of local authority departments with little previous interaction, landowners and business representatives to set broad agendas for development. These can determine the urban design framework within which decisions are made, can propose the idea of a district, or 'quarter', in which living, working, shopping and cultural provision can be intermixed, and question assumptions about land use based on statistical and undifferentiating models (Rowland, 1995);[30] given this, they could equally look to environmentally sympathetic approaches to building design, provision of allotments for growing food, recycling of waste and energy conservation.

Action planning, defined by Nick Wates as 'an approach to planning and urban design involving the organisation of carefully structured, collaborative events at which all sections of the local community work closely with independent specialists from all relevant disciplines' (Wates, 1996: 15) has the benefits of cutting through professional boundaries and mechanistic approaches. It works against a gradual decline in the accountability of civic authorities, and of the public realm in general – Bookchin writes:

> We will want to know what the concepts of 'city' and 'citizenship' really mean – not simply as ideal definitions but as fecund ecological processes that reveal the growth of communities and the individuals who people them – indeed, that turn them into a genuine public sphere and a vital body politic
>
> (Bookchin, 1992: 11)

whilst philosopher Aaren Gare argues that this is possible when a new narrative has taken the place of past grand narratives:

> we only know what to do when we know what story or stories we find ourselves a part of. If this is the case, then to know what to do about the environmental crisis requires the creation of stories which individuals

FIGURE 57 Siah
Armajani, text in
railings at Battery
Park City – is it
still a mad,
extravagant city?

can take up and participate in, which will reveal to them why there are
the problems there are and how they arose, how they can be resolved and
what role individuals can play in resolving them.

(Gare, 1995: 140)

Gare adds that for such new stories to work they must be 'polyphonic' in representing a multiplicity of perspectives,[31] and strong enough to confront those which
are inherited and demonstrate the failure of past (monological) narratives. Perhaps
this is an appropriate point of departure for art, which gives form to the narratives
of a society, being an arena either of conformity, as in the public monument and
conventional public art, or resistance, as in new genre public art. But perhaps the
events of popular culture, such as carnivals and festivals, are at least, or more,
effective in constructing these new, polyphonic stories.[32]

STRATEGIES FOR ART, CRAFT AND DESIGN

Strategies for art, craft and design in response to an agenda for liveable cities take
two main forms, parallel with or contextualised by the practices of people-centred
urban design and action planning. The two forms are the integration of art and
craft work in the design of the built environment, and the intervention of artists

and craftspeople in the public realm – the decorative and the critical, with a possible third area which is simply the presence in cities of artists as creative workers, whether or not they reference the city in their work or locate it in public spaces. It might once have seemed as if there was a dualism of contemporary sculpture sited outdoors and what then seemed a more radical (less individualist) approach which the purveyors of contemporary fine art tended to call 'bland urban design'; today, the colonisation of the street by contemporary art, which increasingly involves artists of international status such as Gormley and Oldenburg, seems off the map of art in the public realm, and is seldom in the public interest, so that a new configuration sets art as urban design beside art as engagement in a model in which the neatness of a single solution is not sought.

Integration

Amongst examples of the integration of art in urban design are Tess Jaray's designs for paving, and (in collaboration with sculptor Tom Lomax) street furniture, in Birmingham's Centenary Square, Wakefield's cathedral precinct and the external spaces of the General Infirmary at Leeds; Siah Armajani's texts set in bridges and railings (as in Battery Park City) (Figure 57); the work of numerous artist-blacksmiths, mosaicists and glass artists; John Maine's designs for bollards and his consultancy in the re-planning of Lewisham; Gordon Young's fish designs set into the pavements of Hull (Figures 58 and 59); seating by Scott Burton and bridges by Richard La Trobe-Bateman; designs for manhole covers in Seattle, and the design of railings for the New York subway by Valerie Jaudon and Laura Bradley. This kind of art may be low-key, yet encourage both a sense of ownership and use of public space if community links and local narratives are developed from the outset, as in the St Peter's Riverside project in Sunderland; it also suggests a renewed history of decoration in which motifs and patterns carry meaning, able to signify what Jaray has described as 'spiritual and symbolic constants' reflecting 'fundamental and continuous human need'; she continues:

> these archetypal patterns also relate closely to natural growth; the order that is perceived in the chaos of the forest . . . seems to be that of that same order that exists when architecture appears at its best as an expression of its society's needs, whether the simple organic geometry of thatched huts . . . the gothic structure of the Marsh Arabs, or the intricate but unified patterns of Seljuk building.
>
> (Jaray, cited by Cork, 1988: 33)

To argue that Jaray's work is an element in the construction of convivial urban environments is not to argue for conventional public art, nor artist–architect collaboration on the basis of the competitive individualism which is part of the

FIGURES 58 AND 59   Gordon Young, *Fish Pavement*, Hull

professional ideology of artists and architects in western societies: urban designer Ian Bentley writes that 'the romantic art tradition of originality through self-expression is utterly bankrupt in the context of the late-capitalist development process [which encourages] designers into vandalising the urban public realm';[33] and artist Dieter Magnus writes that 'Environmental art is . . . related to a certain town planning situation and to the people living there . . . not isolated works of art but forms of integration'.[34]

Whether this relinquishing of artist–hero status characterises the extensive siting of art in the re-design of public spaces in Barcelona in the 1980s is open to discussion. On one hand, the re-design followed contact with neighbourhood groups, and is claimed to represent mixed use development (Mackay, 1991), whilst on the other it signifies a civic intention to be a capital of the western Mediterranean more related to state and corporate interests. Garry Apgar, in an article for *Art in America*, describes Richard Serra's white wall dividing the Placa de la Palmera into planted and open areas for sitting and activity (in collaboration with architects Pedro Barragán and Bernado de Sola) as one of the 'boldest attempts to achieve the ideal of fostering a sense of place and community' (Apgar, 1991: 112), though he also records that 'all but two . . . of the finished projects I visited had been damaged in one way or another' (ibid.: 117), concluding not that this represents a voice denying ownership but that 'the city government will have to admit that art in public places left unprotected – or too easily accessible – is art that will eventually be damaged or destroyed' (ibid.: 118), a response which suggests that public art remains, to this reviewer, museum art outdoors. Similarly, the vandalism which destroyed Vong Phaophanit's *Ash Wall* sited near the Thames Barrage in East London could be discussed in terms either of unsociable behaviour by ungrateful recipients, or of a voice of dissent against the imposition of bourgeois taste.[35] There is, on the other hand, no record of damage to Jaray's street furniture or paving, and the experience of community arts groups is that a sense of ownership is the best guarantee of preservation.

INTERVENTION

Intervention is less comfortable than integration, may provoke outrage, as in the art of AIDS activism, and is essential, including in its rupturing of illusions of consensus, for a democratic society (Mouffe, 1996). Art as intervention in the public realm is a form of continuing social criticism which resists the institutionalisation of conventional public sculpture. Its roots range from the social sculpture of Joseph Beuys to 1960s happenings and the influences of Marxism, feminism and ecology (Lacy, 1995); the strategy involves a redefinition of art as a critical realism which does not record urban experiences but seeks to change them according to ideas of social justice and community, for many artists beginning

with a personal transformation and relinquishing of the notion of the artist as hero, bohemian or victim. This intervention by professionals exists alongside the spontaneous visual interventions of urban dwellers through graffiti and 'unofficial' street murals. Some kinds of art, such as the altering of advertising posters to invert their messages, is between the two.

The projects described in the previous chapter addressing social healing, cultural diversity and environmental awareness are cases of art as intervention, as are the *Conversations in the Castle* curated by Mary Jane Jacob for Atlanta in 1996, which confront the questions 'who is the audience for art' and 'who owns culture' through use of the internet as well as 'front porch' conversations on a deck outside the 'castle' (an early twentieth-century mansion in Atlanta's cultural district). Jacob writes:

> The 'art of public engagement' is misunderstood and often maligned by the mainstream and critical contemporary art fields. It is vital to reassert the presence of community-based art. . . . The field must expand its definition of art by broadening the knowledge of what art means within other societies

(Jacob, 1996)

and that community-based practitioners use a 'many-layered approach' to deseg-regate the meanings of art and life.[36] The *Conversations* project, like *Culture in Action*, locates artists in community settings in which the 'conversation' with members of communities and between artists is the work. Critically referencing the internationalism of the 1996 Olympics (in Atlanta) which is dominated by major powers, the artists are from outside the USA, several from non-western countries; they include Ery Camara from Senegal, Mauricio Dias and Walter Riedweg from Brazil, Navin Rawanchaikul from Thailand, and an artists' collective from Slovenia.

One of the implications of interventionist art, and the point at which it departs from art integrated in urban design, is its expansion of the definition of the location of art, from physical site to public realm, a question initially raised by Phillips in 1988 in her article for *Artforum* (Phillips, 1988). This is a response to the possibilities of mass communication which create an electronic space, and an attempt to focus on the public interest, approaching the framework in which city form is determined rather than the mechanisms through which it is designed. It also transgresses the boundaries of gendered space, as demonstrated by Peggy Diggs' *Domestic Violence Milk Carton* project (Phillips, 1994; 1995b). Just as inter-ventionist art questions the conventions of art practice through a deeper social discourse, so it also resists the structures of power in society and challenges, in its participatory methods, the individualist and isolationist structures of value which, according to Gablik, are leading us to destruction.

Sustainability?

Interventionist art is, by its nature, temporary. Both intervention and integration are able, in differing ways, to contribute to new models of urban dwelling, to new approaches to participation and community. They may need, in order to be sustainable critical and empowering practices, to develop outside current structures of arts management and funding which are closely linked to the structures of power and value of late capitalism and have fostered conventional public art as complicit in socially divisive urban development. That fragmentation is, according to Gablik and Spretnak, a consequence of Cartesian dualism.

What if this dualism could be healed? Then: the integration of art and everyday life; and the end of an art market which depends on a constant expansion and escalation of monetary values (concealed in and explained as taste), one facet of the expanding markets of consumerism which result from a picture of the world as a site for endless exploitation. But the implications are wider, as set out by Gablik in *The Reenchantment of Art*; the ecological values she proposes are also embraced by many alternative communities which have withdrawn from urban life, and include:

awareness of the oneness of humanity and all life . . . a commitment to personal change and social change . . . cooperation and some form of sharing of resources and skills . . . a dedication to healing the earth . . . an emphasis on 'living lightly on the earth' . . . development of some degree of self-sufficiency . . . a commitment to non-violence . . . a commitment to process . . . a dedication to 'thinking globally, acting locally'

(McLaughlin and Davidson, 1985: 23–4)

offering a framework, derived from rejection of the anti-social aspects of urban industrial civilisation, within which methods such as action planning and activist art, at times direct action, may be effective in building, within the ruins of the cities of modernity and postmodernity, a new model of collaborative dwelling, differentiating human want from human need as proposed by Victor Papanek in *Design for the Real World*; that the values of an alternative society have begun to develop largely in rural communities does not mean they are inapplicable to urban situations, any more than the idea of appropriate technology developed in context of foreign aid.

Whilst Serra's *Tilted Arc* epitomises a kind of heaviness – a scarring of the urban fabric, the violence of what Gablik terms the victim–dominator approach (Gablik, 1991: 62–3) – it also represents the complicity of art in the trend to urban dereliction – in its dislocation of a space it treats as value-free, its use of materials requiring considerable investment of non-renewable energy. To make art for 'living lightly on the earth' implies working in sympathy and equity with communities, using materials which can be recycled or are part of the local environment. If some of this art seems tentative or ephemeral, perhaps this is a necessary phase through which to work. The engendering of conviviality may require, also, the space created by an art which simply begins a process of becoming. At the end of *Deschooling Society*, Illich writes:

We now need a name for those who value hope above expectations. We need a name for those who love people more than products . . . who love the earth on which each can meet the other . . . who collaborate with their Promethean brother in the lighting of the fire and the shaping of iron, but who do so to enhance their ability to tend and care and wait upon the other

(Illich, [1971] 1973: 116)

and perhaps this is the point of departure for artists and craftspeople, their work either integrated into the design of cities, or intervening critically in the determination of what a city is, contributing to a way in which urban dwellers, too, may live lightly on an earth which they value and adorn.

NOTES

•

INTRODUCTION

1 There are also established practices of public art under state patronage in Eastern Europe, currently undergoing revision in the light of political change, and in some countries in the southern hemisphere, such as Colombia. These are outside the scope of this book.

2 See Conroy and Litvinoff, 1988, for discussion of sustainable development.

3 See McLaughlin and Davidson, 1985, on alternative communities, and also the discussion in Chapter 9.

4 Wilson, 1991: 156: 'planning is necessary if cities are to survive. What needs to change is the ultimate purpose of planning'.

5 See Gablik, 1991.

6 See Selwood, 1995. A study of the social impact of arts programmes in ten British cities is being conducted at the time of writing (1996–7) by Comedia.

7 The first city in the USA to adopt a Percent for Art Policy was Philadelphia in 1959; for a full list (as at 1987) see Cruikshank and Korza, 1988: 287–95.

8 Identical versions of *Hammering Man*, a 48 feet high steel kinetic sculpture, are in Seattle and Frankfurt.

9 Brian McAvera writes 'the whole point about Gormley's work is not that it is site-specific but that it is site-general. The concept is so vague that it will take the imposition of almost any roughly analogous situation' (McAvera, n.d.: 113).

10 The terminology of the internet – 'rooms' and 'sites' – uses that of space, not time. See also discussion of Laclau's view of space and time in Chapter 2. Its potential for subversion has been tested by Heath Bunting – see *The Guardian*, 11 September 1996, pp.14–15, suggesting that, despite its terminology and the danger that its 'spaces' will be subject to the same kinds of corporate encroachment as public space, it will be a vehicle for activist art, as street murals were in the 1970s and defaced posters more recently.

11 *Travailler* for the International Exhibition and *Les Transports des Forces* for the Palace of Discovery, Paris, 1937.

12 See, for example, Cardiff Bay, 1990; Arts Council, 1991.

13 For critical discussion of the institutions of art and their ideologies, see Pointon, 1994.

In this, Colin Trodd cites Sir Robert Peel's statement to Parliament (Hansard, XIV, 3 July – 14 August 1832, col. 664) in support of a national gallery: 'The rich might have their own pictures, but those who had to obtain their bread by their labour, could not hope for much enjoyment. . . . The erection of the edifice would not only contribute to the cultivation of the arts, but also to the cementing of the bonds of union between the richer and poorer orders of state' (Trodd, 1994: 33); Brandon Taylor, writing on the establishment of the Tate Gallery, notes that 'It comes as no surprise . . . that the cameo illustrations printed in the London daily and weekly magazines between 1892 and 1897 produce Millbank as an educational and leisure site – in one case [illustrated in his text] complete with sailing boats on the Thames' (Taylor, 1994: 21).

14 The exceptions are mainly confined to art from the pre-modern period, such as the fresco paintings made as integral elements of the buildings from which they have been detached and in the process re-coded as 'art'. It could be argued that altarpieces which are portable but depict a local iconology are also conceptually inseparable from their locations.

15 See Chapter 3 for further discussion.

16 For accounts of community participation in urban planning and design see *Urban Design Quarterly*, 49, Jan. 1994, in particular John Thompson, 'Process and Product' (pp.20–22); he concludes: 'It is the participatory process of community architecture and planning that offers our professions the greatest hope of understanding the real nature of both the people and the place, without which our urban futures are likely to be just as inappropriate as our more recent urban pasts'.

17 Willett notes Dean Farrar's sermon to the Liverpool congress of the National Association for the Advancement of Art in 1888: 'to refine, to elevate, to brighten, not only the palace of the noble but the cottage of the poor' (Willett, 1967: 222). The British Medical Journal, 8 June 1895 states: 'The poor are very fond of putting up pictures in their cottages, and must often long to see some pictures on the bare brick walls of the infirmary'.

18 Willett sets out four reasons for new interest in the arts which could be summarised as: the boredom of war; the upheaval of war; optimistic planning for a new Britain from the Beveridge Report; the influence of J. M. Keynes on public policy. This produced the War Arts Advisory Committee and Council for the Encouragement of Music and the Arts; a report, The Arts Enquiry – The Visual Arts, was produced by a committee set up by the Dartington Hall Trustees and chaired by Julian Huxley in 1946. The Arts Council was formed to increase access to fine art and improve the standard of execution of the fine arts. Selwood (1995) gives a detailed account of ideological contradictions within the intentions of the Arts Council and between a romantic idea of the modern artist and social needs.

19 For discussion of the contradiction between modern art's view of the city and its production to a large extent in cities, see Deutsche, 1991a and 1991b.

20 David Hoy writes in *Critical Theory*: 'Theory that is critical is in the first instance critical of itself. Unlike traditional theory, which assumes its own neutrality and therefore neither does nor can investigate itself for blindness and bias, critical theory would suspect itself of both' (Hoy and McCarthy, 1994: 105).

21 Lang and his contributors draw a parallel between Los Angeles after the riots and

Sarajevo after the Balkan civil war, and cite statistics of crime, violence and dereliction.

1 THE CITY

1 De Certeau argues that 'The desire to see the city preceded the means of satisfying it. Medieval and Renaissance painters represented the city as seen in a perspective that no eye had enjoyed' (de Certeau, [1984] 1988: 92), and Wilson that Alberti's treatise on architecture and planning (1434) was 'the model for many others – and even for the whole concept of planning' (Wilson, 1991: 19).

2 This breaks from the pre-modern practice of representing value by scale – as in religious paintings in which the most important (divine) figures are largest and the least (human) smallest.

3 J. B. Jackson, in *American Space*, writes of the 'monotony of the façades, together with the dreary predictability of the street numbers . . . a city without variety or colour . . . the grid of uniform streets, the levelling of every hill' but argues that the unplanned processes of urban expansion created a diversity out of this – 'evolving innumerable specialized areas, some no larger than a block or two' (Jackson, 1972: 205). The claims for planning and ad hoc development, as means to achieve liveable cities, remain contended – see Wilson, 1991: 135–59.

4 One of the first re-configurations of a city was Sixtus V's transformation of Rome from 1585 as a sequence of vistas. Richard Sennett writes, in *The Conscience of the Eye*: 'Recourse to perspective as a model for urban design suggested . . . a new way to establish the meaning of a street line leading to a center' and '[Sixtus] wanted to connect these sites [of pilgrimage] through straight streets that established tunnels of vision' (Sennett, [1990] 1992: 153).

5 Berman adds that 'If we want to locate Faustian visions and designs in the aged Goethe's time, the place to look is not in the economic and social realities of that age but in its radical and utopian dreams; and, moreover, not in the capitalism of that age, but in its socialism'. He notes that amongst Goethe's reading in the 1820s was the Paris paper *Le Globe*, a voice for the Saint-Simonists. But utopian visions, like social realities, are socially produced.

6 'The ragpickers were one of the most abject and notorious groups in Parisian society. They lived like nomads in shanty towns' (Wilson, 1991: 54).

7 Rilke arrived in Paris in 1902; his fictional impoverished Danish poet living in a world of interiority amidst the city of modernity is somewhat overlooked in comparison with the attention given, largely through Benjamin's writing, to Baudelaire's *flaneur*. Rilke may comment less directly on metropolitan sociation, but his 'notebooks' include, for example, this image of the neighbour: 'a creature that is perfectly harmless; when it passes before your eyes, you hardly notice it and immediately forget it again. But as soon as it somehow, invisibly, gets into your ears, it begins to develop, it hatches, and cases have been known where is has penetrated into the brain . . . I have had innumerable neighbours' (Rilke, [1910] 1988: 168).

8 T. S. Eliot's *The Wasteland* could be discussed using a similar framework.

9 See Wilson, 1991: 89 for the role of the feminine in *Metropolis*.

10 The roots of many alternative communities which still flourish are in the 1960s, but the solution they constructed is often based on the self-sufficiency of small rural units. The energy of this movement was, is, largely not applied to urban problems.

11 'In modern Western culture . . . The "oral" is that which does not contribute to progress; reciprocally, the "scriptural" is that which separates itself from the magical world of voices and tradition' (de Certeau, 1984: 134).

12 Edward Soja opens a discussion of Orange County, California: 'You can have anything you want in Orange County, where every day seems just like yesterday . . . every place is off-center, breathlessly on the edge, but always right in the middle of things' (Soja, 1992: 94).

13 'The recovery of the essential activities and values that first were incorporated in the ancient cities, above all those of Greece, is accordingly a primary condition for the further development of the city in our time' (Mumford, [1961] 1966: 648); Lefebvre sees Mumford as an idealist who considers cities made up of free men regardless of the division of labour and class struggles (Lefebvre, 1996: 97).

14 For example, the image of the archaic societies of the Near-East, such as Babylon, in French nineteenth-century *salon* painting, or the work of the Orientalists, compensates for bourgeois conventions of prudery with displays of sexual licentiousness.

15 Wilson argues that Mumford's dislike of metropolitan cities, like that of William Morris, was founded in the escape from patriarchal control provided for women by the metropolis (Wilson, 1991: 18).

16 Sibley cites Bronislaw Geremek's *The Margins of Society in Late Medieval Paris* (1987) on the segregation of spatial zones for vice.

17 The analogy between modern planning and art is made by Siegfried Giedion, who likened modern highways to cubist paintings in his Harvard lectures of the 1930s (Berman, 1983: 302).

18 There are exceptions: Sparta, for instance, had no city walls. Sennett cites Rykwert's case that the Egyptian sign for a city was a cross in a circle: 'This hierograph . . . suggests two of the simplest, most enduring urban images. The circle is a single, unbroken closed line: it suggests enclosure, a wall or a space like a town square; within this enclosure, life unfolds. The cross is the simplest form of distinct compound lines; it is perhaps the most ancient object of environmental process . . . Crossed lines represent an elemental way of making streets within the boundary, through making grids' (Sennett, [1990] 1992: 46–7).

19 Athens packed bodies together in two kinds of spaces, each of which gave the crowd a distinct experience of spoken language. In the agora, many activities happened at once, people moving about, speaking in little knots about different things at the same time. . . . In the theatres of the ancient city, people sat still and listened to a single, sustained voice' (Sennett, 1994: 52). At a discussion on this book (at the ICA, London, 13 March 1995) Sennett emphasised the loss of sensual experience for modern city dwellers whose experiences are mediated by the enclosure of the automobile, being thus only visual and rapid.

20 'This was not simply a question of convenient access to the marketplace where artisans

sold their wares. Like the agora itself, the workshops, recognised meeting places for discussion, were an important part of the public realm' (McEwen, 1993: 75).

21 'The agora [of Athens] is both a commercial and a political centre; only through his relation to the agora can Hermes the god of commerce make contact with the cult of the Twelve Gods [of Olympus], the expression of political unity' (Brown, 1990: 107–8).

22 McEwen sees Anaximander as 'the first real philosopher', his work 'the watershed in the transition from myth to philosophy, the transition whereby a so-called rational account of the world takes the place of a so-called irrational one' (McEwen, 1993: 9). Her conclusion is that architecture is not the representation of metaphysics, but that philosophy 'was first grounded in architecture'. For a different reading of the fragment, see Heidegger, [1950] 1975.

23 'The archaic polis was an uncertain place that needed to be anchored at the strategic points of centre, middle ground, and outer limit by the new sanctuaries. It was not a vessel with a fixed form, but, like the appearing surface of a woven cloth ... had continually to be mended or made to reappear' (McEwen, 1993: 83–4).

24 'Weaving ... "consists of the interlacing at right angles by one series of filaments ...". Harmonia, close-fitting, can be a feature of the tightly woven cloth only: a textile with a loose weave is not, so to speak, "harmonious". It does not, properly speaking, appear at all' (McEwen, 1993: 83).

25 Sibley uses Swift's *Gulliver's Travels* to illustrate the point: 'Consciousness of pollution in Lilliput is heightened by the geometry of the landscape. In particular, the metropolis, Mildendo, had a highly ordered design with strong internal boundaries and the populace was excluded from the centre – a sacred space, the home of the emperor' (Sibley, 1995: 54).

26 'Haussmann's slum "clearance" simply broke up the working class neighbourhoods and moved the eyesores and health hazards of poverty out of central Paris and into the suburbs ... behind the scenes his building projects initiated a boom of real estate speculation' (Buck-Morss, 1995: 89).

27 'It is common knowledge that the seventeenth century created enormous houses of confinement; it is less commonly known that more than one out of every hundred inhabitants of Paris found themselves confined there, within several months' (Foucault, 1967: 38).

28 'Drains for the disposal of urine be kept well away from the walls, as the heat of the sun may corrupt and infect them very much' (Alberti, *On the Art of Building*, IV, p.114, cited in Colomina, 1992: 344).

29 In *The Uses of Disorder*, Sennett draws on the developmental psychology of Erik Erikson to construct an image of the rigid self which generates a heightened desire for order in face of external confusion, inhibiting the development of the self through overcoming the crises of selfhood.

30 'The boundary between the inner (pure) self and the outer (defiled) self, which is initially manifest in a distaste for bodily residues but then assumes a much wider cultural significance' (Sibley, 1995: 7).

31 For discussion of Simmel, see Frisby, 1992; Smith, 1980, Chapter 3; and Savage and

Warde, 1993, Chapter 5. Savage and Warde claim (p.111) that Smith misunderstands Simmel in linking his *Metropolis and Mental Life* too closely to Wirth's *Urbanism as a Way of Life*, 1938: 'Wirth misunderstood Simmel's essay in important respects, for his project was rather different. Simmel was primarily concerned to establish that urban culture was the culture of modernity.'

32 For a full list of Benjamin's categories, which include things as well as types, see Buck-Morss, 1995: 50–1.

33 'The urge to simplify in order to generalize is drawn from the practice of the natural sciences. The exponents of the relatively new discipline of sociology were keen to model their work upon the natural sciences and thereby make legitimate their discipline' (Duncan, 1996: 257).

34 Geographer James Duncan states that, following Burgess, researchers 'devoted careers to searching for concentric zones all over the world' (Duncan, 1996: 259).

35 Chicago was already affected by ideal notions of the city, its lake front developed according to the *Plan for Chicago* by Daniel Burnham (1909), based on the ideas of Haussmann and Louis XIV (Wilson, 1991: 69). The ring model does not replicate this plan, and is a further stage of abstraction, in which the demands of capital replace those of the state.

36 'The ability to control emanates in large part from nodality/centrality itself' (Soja, 1989: 153).

37 Soja argues that this differentiation of urban from rural is 'not inaccurate' but conceals a more fundamental aspect: 'Cities are specialized nodal agglomerations built around the instrumental "presence availability" of social power. They are control centres' (Soja, 1989: 153).

38 For discussion of the differences between Wirth and Simmel, see Savage and Warde, 1993: 110–14. They state four differences and conclude that 'In all these ways Wirth's innovations proved unhelpful' (p.114).

39 Urban historian Jerry White's study of Campbell Road in north London demonstrates that perceptions of a neighbourhood differ according to the viewpoint of the observer: from its middle-class borders, Campbell Road was known as 'the worst street in north London' whilst for its inhabitants its spirit was 'represented as a fierce and self-respecting exclusiveness [which] assumed and demanded an internal loyalty, a cohesion in the face of the outside world' (White, 1981).

40 See discussion of Lefebvre in Chapter 2.

2 SPACE, REPRESENTATION AND GENDER

1 See Forester, 1987a, part IV

2 Gemäldegalerie, Staatliches Museum, Berlin, used as cover illustration on the 1991 Penguin edition of Mumford's *The City in History*.

3 'Beyond those . . . abstract, one-dimensional indications, we encounter the space of the vibrant, everyday world . . . the gendered city, the city of ethnicities, the territories of different social groups, shifting centres and peripheries' (Chambers, 1993: 188)

4 Harvey notes the importance of the production of registers of land ownership in the

abstraction of space: 'Cadastral survey permitted the unambiguous definition of property rights in land. Space came to be represented, like time and value, as abstract, objective, homogenous, and universal in its qualities' (Harvey, 1989: 176–7).

5 Sometimes the correspondence has a life of its own. Phillips writes of a seven-year-old child who thought the equator was a 'real' line round the world (Phillips, 1989).

6 See de Certeau, 1988: 91f.: 'The gigantic mass is immobilized before the eyes. It is transformed into a texturology in which extremes coincide – extremes of ambition and degradation, brutal oppositions of races and styles . . .'.

7 'In place of the transcendental comparison between the image and perceptual private worlds, stand the socially generated codes of recognition' (Bryson, Holly and Moxey, 1991: 65).

8 James Duncan cites this passage and comments 'This sounds rather like Burgess' conception of the city. But Barthes does not attempt realist representation, instead he deploys irony in contrasting the empty centre of Tokyo to the full western centre' (Duncan, 1996: 263). Lefebvre writes: 'Logical relationships . . . may be represented by geometric figures; thus circles, larger ones including smaller, may serve to symbolise concepts. Such representation merely illustrates relations which have no basic need of representation, since they are themselves of a strictly formal nature' (Lefebvre, 1991: 293–4).

9 James Duncan writes: 'What is it about then? The answer is that it is about writing, or more precisely about Barthes' writing. The emptiness and the fragmentary nature of the Tokyo he describes is like his own theoretical project' (Duncan, 1996: 264).

10 This model is also one of refusal to privilege concept over form, or mine for an underlying concept beneath a material form.

11 'Barthes through irony permanently defers [meaning] into the endless empty circles of the poststructural crisis of representation' (James Duncan, 1996: 265); Chambers argues, however, that Barthes' 'shimmer of the signifier' does not necessarily mean 'joining Baudrillard and the cultural pessimists in announcing the end of meaning' (Chambers, 1993: 195).

12 'What do cities mean to people now living in them? Crime, dirt, congestion, expense, hustle, danger, noise, aggravation, high blood pressure, unwanted violations . . .' (Roszak, 1973: 417).

13 See Sennett, 1990: 46.

14 'Allegorical figures representing abstract concepts become such containers, composed of strong, firm materials . . . Cities often wear a full panoply too, in the consular diptychs of the Roman empire and Byzantium, in which Rome, Constantinople or Antioch appear. The emblem of the city walls, represented by the turreted headdress, is the "very particular . . . Quoiffure" of personified cities' (Warner, 1987: 258–60).

15 'Paris was the city sexualized. Poets sometimes likened Paris to a prostitute, but more often sang her praises as a queen' (Wilson, 1991: 47).

16 Lefebvre continues: 'The ultimate effect of descriptions of this kind is either that everything becomes indistinguishable or else that rifts occur between the conceived, the perceived and the directly lived – between representations of space and representational spaces. The true theoretical problem, however, is to relate these spheres to one another,

and to uncover the mediations between them' (Lefebvre, 1991: 298).

17 Janet Wolff draws attention to the absence of the 'flaneuse' in modernity. See Wolff, 1989.

18 For an application of these categories to the Roman state, see Lefebvre, 1991: 245.

19 'Out of this process emerged, then, a new representation of space; the visual perspective shown in the works of painters and given form first by architects and later by geometers' (Lefebvre, 1991: 79) and 'The desire to see the city preceded the means of satisfying it. Medieval or Renaissance painters represented the city as seen in perspective that no eye had yet enjoyed' (de Certeau, [1984] 1988: 92).

20 *Della Pittura* (1435) should be seen in context of Alberti's other works – *Della Famiglia*, trans. R. N. Watkins [The Family in Renaissance Florence], Univ. S. Carolina Press, 1969, and *On the Art of Building* in ten books, trans. J. Rykwert, N. Leach and R. Tavernor, MIT, 1988.

21 Wilson notes that Alberti was, as a priest, unmarried (Wilson, 1991: 15).

22 'The first truly private space was the man's study, a small locked room off his bedroom which no one else ever enters, an intellectual space beyond that of sexuality. Such rooms emerged in the fourteenth century and gradually became a commonplace in the fifteenth century' (Wigley, 1992: 347).

23 'The weather being cold, he got into a stove in the morning, and stayed there all day meditating; by his own account, his philosophy was half finished when he came out . . .'. In a footnote Russell questions this but admits 'Those who know old-fashioned Bavarian houses, however, assure me that it is entirely credible' (Russell, [1946] 1961: 542).

24 'The first point is to see the persons with the eyes of the imagination . . . The second is to hear what they are saying . . . The third is to smell and taste the infinite sweetness and delight of the Divinity . . . The fourth is to feel with the touch . . .' (St Ignatius Loyola cited in Martz, 1962: 78). However, in other sources, such as (attrib.) Lorenzo Scupoli, *The Spiritual Combat*, which appeared in a 2nd edition as *The Spiritual Conflict*, Rouen, 1613, 'the rational faculty is placed between the divine will above and the sensitive appetite below, and it is attacked from both sides' (ibid.: 127).

25 'Not only is it now possible and necessary to narrate the outer world from an inner place, by means of a clarified and transparent instrumental language, and similarly to reflect on others as Other, but – more insidiously – the subject can, and now must, reflect on itself in the same fashion' (Barker, 1984: 53).

26 Barker notes the case of Caspar Barlaeus, a figure in Rembrandt's *Anatomy Lesson*: 'a leading intellectual and noted neurotic, who wrote poetry in praise of Tulp's dissection . . . and dared not sit down for fear that his buttocks, which were made of glass, would shatter' (Barker, 1984: 115).

27 The easy, reasonable tone given throughout to the conversational persona who utters the Discourse is essentially a device of seduction . . . A guileful democracy of gesture characterises the text's encoding of its own status, feigning to disavow technical expertise or superior ability, always assuming a genial community of understanding between it and its singular reader' (Barker, 1984: 54).

28 See also Xenophon, *Oeconomicus*, trans. H. G. Daykens in *The Works of Xenophon*,

Macmillan, 1897, cited by Wigley, 1992: 334 – 'The gods made provision from the first by shaping, as it seems to me the woman's nature for indoor and the man's for outdoor occupations'.

29 The list could be extended to include abstract expressionism in the form of Willem de Kooning's violent disjunctions of images of women, and new figuration in Lucien Freud's treatment of female flesh in the same vocabulary as soiled lino.

30 A poster designed originally for Public Art Fund, New York, but rejected as unclear (!), the poster was eventually sited on buses. Its subsidiary text states 'Less than 5% of the artists in the Modern Art Sections are women, but 85% of the nudes are female.'

31 Of 73 works in downtown Seattle (listed in Rupp, 1991, sections 3–5), 54 (74 per cent) are by men, 13 (18 per cent) by women and 6 (8 per cent) are collaborations involving men and women artists. Of the illustrations in Petherbridge, 1987, 127 (77 per cent) are of work by men, 28 (17 per cent) by women, and 11 (6 per cent) collaborations; those in Jones, 1992, 50 (57 per cent) are of work by men, 26 (30 per cent) by women and 11 (13 per cent) of collaborative efforts (excluding a small number of indeterminate attribution); and those in Heath, 1992, 27 (53 per cent) are by men, 12 (23.5 per cent) by women and the same proportion by collaborative teams or unattributed, excluding illustrations in the chapter by Patrick Nuttgens on landscape.

32 'Public art in the Eurocentric cultures has served the value systems and the purposes of an unbroken history of patriarchal dominance that has despoiled the earth . . . ' (Jo Hanson, cited in Lacy, 1995: 33).

33 'Another problem with the conception of place which derives from Heidegger is that it seems to require the drawing of boundaries. . . . It is yet another form of the construction of a counterposition between us and them' (Massey, 1993: 64).

34 'When geographers gaze at social space . . . their claim to know and to understand rests on a notion of space as completely transparent, unmediated and therefore utterly knowable' (Rose, 1993: 70).

35 Irigaray writes elsewhere of the problem of gender in language and the assumption of universality by the masculine; see, for instance, 'The Unconscious Translation of Gender into Discourse', in Irigaray, 1993: 172f.: 'in French the masculine gender always carries the day syntactically: a crowd of a thousand persons, nine hundred and ninety-nine women and one man will be referred to as a masculine plural . . .'.

36 Christine Battersby takes the proposal literally, and argues that 'Even bearing in mind that Irigaray is writing in a Catholic culture, this treatment of art as mere propaganda cannot be excused' (Battersby, 1995: 131), which seems a case of the intolerance and literalism characteristic of masculine culture.

37 Yet if Kruger, Holtzer *et al.* inherit the conceptual critique of the given parameters of art production and reception, they do so not uncritically. . . . these . . . artists have opened up the conceptual critique of the art institution in order to intervene in ideological representations and languages of everyday life' (Foster, 1985: 100).

38 Pollock acknowledges Edward Said's *Orientalism* as a key source for the deconstruction of colonial attitudes to culture.

39 'Within neocolonial white-supremacist capitalist patriarchy, the black male body

continues to be perceived as the embodiment of bestial, violent, penis-as-weapon hyper-masculine assertion. Psychohistories of white racism have always called attention to the tension between the construction of black male body as danger and the under-lying eroticization of that threat that always then imagines that body as a location for transgressive pleasure. . . . a process that takes place primarily in an aesthetic realm . . .' (hooks, 1995: 205).

40 Of the artists given brief biographies in (ed.) Suzanne Lacy, 1995, *Mapping the Terrain*, 36 (46 per cent) are women, 32 (41 per cent) men, and 10 (13 per cent) collaborations; of the contributors to the book, 9 are women and 3 men.

41 'Adorno's meditations on the social implications of Auschwitz led him to the belief that any idea of harmonizing with the world . . . is cheap optimism . . . there is no meaningful order now . . . to which anyone can belong' (Gablik, 1991: 31).

42 'Medical discourses were especially important to the development of these arguments. In the eighteenth century, medical men searched for the natural laws which they thought structured women's physiology and thus their psychology, and by the mid-nineteenth century women's spontaneous ovulation and maternal instinct had been "discovered"' (Rose, 1993: 73).

3 THE MONUMENT

1 The monument also plays a key role in colonialism, both in the colonised lands and in the 'home' state, legitimising oppression by subsuming it, and the activities of opportunistic colonisers, in the myth of liberalism.

2 Jon Bird argues, citing Eric Hobsbawm, that the period 1870–1914 is one in which traditions were created for the redefinition of social relations and the central role of the state. 'The period witnessed the management of the population through the recog-nition of bonds of familiarity and allegiance to the nation, the community, the past' (Bird, 1988: 30). It could be noted that many statues and memorials in this period were funded by public subscription, though this may simply suggest a public persuaded by the 'official culture' used at the time to establish a national identity.

3 Illustrated in Tolstoy, Bibikova and Cooke, Plates 21 and 22.

4 For example, a statue of Cecil Rhodes in a suit, accompanied by Edward VII and George V, at the High Street entrance of Oriel College, Oxford (Darke, 1991: 156).

5 Maraise's argument reflects, although there is no evidence that it is derived from the transcendent quality of abstraction in the paintings (*c.* 1900–14) of Wassily Kandinsky and Franz Marc.

6 Walter Benjamin, *Gesämmelte Schriften*, V, p.188, cited in Buck-Morss, 1995: 89.

7 French is better-known for the *Lincoln Memorial* in Washington; for discussion of the monuments on the Mall in Washington see Griswold, 1992: 79–112.

8 *Sir Henry Tate* by Thomas Brock, Brixton Hill, London, unveiled 1905 (Darke, 1991: 72). A student at the University of Portsmouth reported hostility from a local person when photographing the bust. There has also been an exchange of letters in *The Guardian*, Ibrahim Thompson drawing attention to the Tate's link to slavery, and curator Simon Wilson responding that slavery was abolished in 1833 whilst Tate built

his first refinery in 1873. (*The Guardian*, 8 and 15 February 1996.)

9 Holub sees a parallel between Gramsci's work and that of the Frankfurt School (Holub, 1992: 12); she accepts that Gramsci never met the Frankfurt School theorists and that they probably did not read each other's work.

10 Edward Said argues that what was known of Egypt by imperial nations was the product of a reconstruction through drawings: 'First the temples and palaces were reproduced in an orientation and perspective that staged the actuality of ancient Egypt as reflected through the imperial eye; then ... they had to be made to speak [through Champollion's decipherment of scripts] ... then, finally, they could be dislodged from their context and transported to Europe for use there' (Said, 1994: 141–3).

11 See Warner, 1987, Chapter 4.

12 A wall-painting in the Hotel de Ville, Paris, by E. Duez, illustrated in Warner, 1987, Figure 18. There is also an allegory of Gas by A. Itasse.

13 *Transportation*, 1912–14, designed by Jules-Félix Coutan, carved in limestone blocks by John Donelly.

14 The *Cable Street Mural*, London, by Desmond Rochfort, Ray Walker and Paul Butler, 1983, was painted at a time when parties of the extreme right had begun to receive increased support in local elections; the mural has been paint-bombed by members of such parties more than once. Rochfort and Butler use a similar device to dehumanise the police in their mural Labour in History: *Struggles for Power* (1986), at the TUC Education Centre, London.

15 al-Khalil (1991: 53, Figure 29) shows a giant plywood portrait of Husain behind a model of the Babylonian gate of Ishtar. In the caption to an equestrian statue of Faisal I (Fig. 71) re-erected by the Ba'athist government in 1989, al-Khali writes: 'The Ba'ath today exude an air of timeless inevitability, a sense of connectedness with the past so profound that its terminus in their rule seems almost logical' (p.131).

16 Saum Song Bo, 1885, 'A Chinese View of the Statue of Liberty', in *American Missionary* 39, cited in Dillon and Kotler, 1994: 'The word liberty makes me think of the fact that this country is the land of liberty for men of all nations except the Chinese. I consider it as an insult ... to call on us to contribute toward building ... a pedestal for a statue of Liberty'.

17 Cincinnati is known for conservative cultural positions, including the banning of X-certificate films such as *Last Tango in Paris*.

18 For discussion of this, from Ernesto Laclau's *Reflections on the Revolution of our Time*, see Massey, 1994: 249–72.

19 Unpublished remarks made by Rodin to a newspaper reporter, cited in Elsen, 1985: 106.

20 Another proposal was Jules Dalou's 100–feet high column, like an obelisk with a round top, or phallus, conceived in 1888 (illustrated in Willett, 1984: 8).

21 Snooks belonged to a local doctor, to whom the monument is dedicated, and wore on his collar the text 'please do not throw stones at this dog'; the representation of a man by his dog seems rather English, and adds a new dimension to a critique of the monument, though within the terms of a national identity.

22 Atherton still regards *Platforms Piece* as successful, regardless of the state of the station

(letter to the writer, April 1996).

23 He cites R. Campbell's 'An Emotive Place Apart', *AIA Journal*, 72, May 1983, p.151, in which Lin is quoted: 'I thought about what death is, what loss is . . . a sharp pain that lessens with time, but can never quite heal over. A scar' (Griswold, 1992: 106).

4 THE CONTRADICTIONS OF PUBLIC ART

1 See, for instances, Townsend, 1984; Fleming and von Tscharner, 1987; Petherbridge, 1987; and Beardsley, 1989.

2 See Selwood, 1995, for both a statement of the lack of evaluative research and the first serious attempt to provide it.

3 The work of Martha Rosler, for example, moves between gallery and street. See Rosler, 1991 and 1994.

4 By the art establishment is intended the informal but structured network of curators, arts administrators, collectors, critics and dealers who between them determine the manufacture of reputations in the artworld. This would be an intriguing place in which to observe the operation of hegemony, for example, in the way private capital uses public institutions to increase the value of investment art, as when a collection of paintings by Julian Schnabel was loaned by the Saatchi collection to the Tate Gallery, thus increasing the work's market value. See Fuller, 1985: 74.

5 'Potentially, any exhibition venue is a public sphere and, conversely, the location of artworks outside privately owned galleries . . . hardly guarantees that they will address a public' (Deutsche, 1991a: 167).

6 Peter Fuller writes: 'These are the skilled, brilliantly controlled and orchestrated paintings of a desperate professional painter who has nothing to say, and no way of saying it' (Fuller, 1980: 102).

7 See Grunenberg, 1994, on the relation of the Museum of Modern Art, New York, to ideology and puritanism.

8 See Lefebvre, 1991: 361, cited in Chapter 2.

9 For instance, the series of talks at the ICA, London, in the Winter of 1995 and Spring of 1996 which included contributions from, amongst others, Richard Sennett, Edward Soja, Beatriz Colomina, Dolores Hayden and Mark Wigley.

10 In the selected bibliography in *Art for Architecture* (Petherbridge, 1987), the earliest entry dates to 1967; *Earthworks and Beyond* (Beardsley, 1989), lists in the general bibliography only five references dated before 1970; *Art in the Public Interest* (Raven, [1989] 1993) includes nine titles prior to 1970 in nearly one hundred books, and the twelve contributors to *But is it Art?* (Felshin, 1995) give only a handful of references to works prior to 1980.

11 Whilst contemporary art is covered by several specialist journals, there is only one in the USA – *Public Art Review* – and none in the UK dedicated to 'public art', although *Stroll* was intermittently published in New York in the 1980s covering street art. For journals covering the 'mainstream', there has been little incentive to give space to a field from which no gallery advertising will result. There are some publications in Europe, such as *Orte* in Germany (recently closed) and *Environnemental* in Belgium, and

L'Institut pour l'Art et la Ville in Givors, France, publishes a series of *Cahiers*.

12 Perhaps the inception of the literature should be dated to the review *L'art public*, published quarterly in Belgium from 1907 following congresses on the subject in 1898, 1900 and 1905. Victor Horta and the socialist deputy Jules Destrée were on its committee, and its remit extended to street furniture as well as monuments.

13 Lacy dates the inception of institutional public art to 1967: 'the contemporary activity in public art dates from the establishment of the Art in Public Places Program at the National Endowment for the Arts in 1967' (Lacy, 1995: 21). In the UK, Willett's report is itself a landmark, recommending the appointment of a public art adviser for Liverpool; 1967 is also the date of the first Black Pride mural in Chicago, and the late 1960s a time when the commodification of art by its market began to be challenged through non-object art forms, such as 'happenings'.

14 Willett's surveys involve a fairly random set of groups, including visitors to the Walker Art Gallery and a local school.

15 The pioneering work of Tim Clark on Courbet, Baudelaire and the 1848 Revolution which located art as socially produced was published in 1973; Linda Nochlin's essay on 'Why have there been no great women artists?' appeared in the same year.

16 Foucault's *Discipline and Punish* was published (in English) in 1977, the same year as Barthes' *Image-Music-Text*; Janet Wolff's *The Social Production of Art* was published in 1981, and in 1984 Griselda Pollock wrote an article, 'The history and position of the contemporary woman artist', for *Aspects* and Suzi Gablik's *Has Modernism Failed?* was published.

17 Willett seems ambivalent on this, later writing that people no longer wish to receive the new vision from 'some intermediate cloud peopled by architects, planners and official committees' (Willett, 1984: 13).

18 'Those who buy art on behalf of the public have an obligation to secure the best art' and 'people who know what is currently available and who have a visual flair (such as modern art museum/gallery curators and visual arts officers) will advise on good decorative art' (Townsend, 1984: 41–2).

19 The absence of commonalities in visual culture was a frequent theme in Peter Fuller's art criticism around this time. In 'Where was the Art of the Seventies', he writes: 'at the moment there are no "given" pictorial conventions which are valid for anything other than small, particular publics' (Fuller, 1980).

20 To extend Jencks' position: would Rembrandt's painting *The Jewish Bride* be interpreted as a symbolic representation of the cosmology of the Netherlands in the seventeenth century (accepting the religious complexities this would involve, between Protestantism, Catholicism and Judaism), or as an expression of tenderness derived from (secular) everyday life?

21 This project will be discussed in Chapter 5. Pelli states: 'Close to 100 percent of the creative work was done when we worked together. In between, the decisions that we had made . . . would be translated into accurate drawings and a new model' (Harris, 1984: 37).

22 For a record of the GLC's support of arts in the community, see *Campaign for a Popular Culture* (GLC, 1986).

23 See Chapter 5 for a discussion of Myerscough in relation to arts policy.

24 The case for commissioning public art as set out in *Percent for Art: a review* is given in the following chapter.

25 A particular problem is the lack of a tested methodology for either qualitative or quantitative evaluation, though this could be developed from the skills of market researchers; more important is the definition of what constitutes a benefit to society.

26 In an earlier text, first published in 1989, Lacy argues that 'Discussions about art in public places are held largely by a class of critics and artists who ignore aspects of a complex and often highly politicised heritage. Questions of aesthetics, largely sculptural in nature, dominate the critical dialogue, with community involvement evaluated as an appendage of, rather than integral to, the critique' (Lacy, 1993: 289).

27 These items could, of course, be critiqued in design history.

28 Lacy uses the phrase 'society of the spectacle' but does not refer directly to Guy Debord's *Society of the Spectacle* (Debord, 1967) which states (paragraph 4) 'The spectacle is not a collection of images, but a social relation among people, mediated by images'; this is related to the capitalist mode of production and resulting alienation, the spectacle becoming (paragraph 21).

29 Bay Press also published critical writing by (and a collection edited by) Hal Foster, and a series of *Discussion in Contemporary Culture* – See for example Foster, 1987; Kruger and Mariani, 1989.

30 Gablik's 1984 *Has Modernism Failed?* was reprinted in the same year, 1991.

31 'Site-specific works do not automatically disrupt our notion of context, and alternative spaces seem nearly the norm' (Foster, 1985: 25).

5 ART IN URBAN DEVELOPMENT

1 Francis Gomila, as town artist, worked with local shop-keepers to create a unified but individualised set of decorative façades along the High Street, within a renovation scheme carried out by Sandwell & Dudley District Council. See Miles, 1989: 123–5 and (for interviews with shop-keepers) Selwood, 1995: 287–8. Selwood reports a positive response from the local authority: 'planners regarded both the enveloping scheme and the decorative scheme as successes' but from a random sample in the street 'only 20 percent thought that the decorations improved the appearance of the shops', although the comments were intermingled with a racist reaction to the predominance of Asian shops.

2 For details of schemes in the USA, see Cruikshank and Korza, 1988. For information on the take-up of schemes in the UK, see Roberts *et al.*, 1993. For information on the public art policies and practices of members of the EC, see *Environnemental*, 6–9: 1993.

3 Sibley cites the example of car advertising in which 'the car [is] a protective capsule which insulates the owner from the hazards of an outside world populated by various "others"' (Sibley, 1995: 63). He also notes that strong spatial boundaries tend to make abjection more likely – 'the strongly classified environment is one where abjection is most likely to be experienced' (Sibley, 1995: 80).

4 See Featherstone, 1995: 96 'the growing number of open-air and indoor industrial or

everyday life museums . . . the physical reconstruction of past localities where preservation of the real merges with simulations', and 'Such postmodern spaces could be regarded as commemorative ritual devices which reinforce . . . a lost sense of place.'

5 Berman recalls its development from an area scheduled for demolition to make space for Robert Moses' Lower Manhattan Expressway to an area of low-rent lofts in the 1970s – 'thousands of artists moved in and, within a few years, turned this anonymous space into the world's leading centre for the production of art' (Berman, 1983: 337).

6 Sibley writes of 'schemes to re-shape the city . . . as a process of purification, designed to exclude groups variously identified as polluting – the poor, in general, the residual working class, racial minorities, prostitutes, and so on' (Sibley, 1995: 57) and gives Haussmann's Paris as a case, also citing Illich: 'The effort to deodorize the utopian city space should be seen as an aspect of the architectural effort to clear city space for the construction of a modern capital. It can be interpreted as the repression of smelly persons who unite their separate auras to create a smelly crowd' (Illich, 1986: 53, cited in Sibley, 1995: 57).

7 Ambrose reminds us that bankruptcy is a protective device which shifts the burden of debt to banks, who presumably recoup it in charges to customers (Ambrose, 1994: 10). Since the much-publicised collapse, most of the office space at Canary Wharf has been rented.

8 See Bloxham, 1995, on alternative regeneration projects: 'The (not so) secret of their successes is bringing new mixed uses into the urban core – uses such as specialist retail and residential accommodation, office and studio work space, bars, clubs, cafes and restaurants . . . these new mixed uses transform urban wasteland into thriving centres of culture, life and commerce' (p.10). What Bloxham does not address (though he cites Covent Garden) is the problem of gentrification. On the other hand, he states that the cost of job creation in one scheme in Manchester was 'a few hundred pounds per job' compared with many thousands in other more standard cases.

9 See Arts Council, 1989: 8–9. Newcastle Arts Centre occupies a city-centre site, and involved the restoration of several hitherto derelict eighteenth- and nineteenth-century buildings, some of architectural interest. The Centre includes a performance space, craft workshops and shops along the street frontage.

10 In London there is a concentration of such schemes, mostly in the east and south-east of the city, co-ordinated by specialist organisations such as SPACE.

11 A text from a contemporary visual art journal, *Frieze*, confirms this in terms of artistic autonomy: 'Artists have a responsibility to art not to anything or anyone else' (Batchelor, 1995). Gablik cites the artist Georg Baselitz: 'The artist is not responsible to anyone. His social role is asocial; his only responsibility consists in an attitude to the work he does' (Gablik, 1991: 61).

12 For example, Susan Tebby was commissioned to design internal courtyards for Quarry House, for the Department of Health and Social Security, in Leeds. The courtyards are not open to public access; the building is known locally as 'the Kremlin'.

13 A report from the University of Westminster shows that local authorities commission around three times as much public art as property developers (Roberts *et al.*, 1993).

14 The opinion of Robert Carnwath QC (to the steering group) was: 'the promotion of

art is not a proper function of planning control' (Arts Council, 1990: 5,1). Developers could therefore be only encouraged to provide art, for example within landscaping (which might properly be required). Another legal point was the 'absolute prohibition in relation to the financial methods of Percent for Art ... There is no power under existing legislation to insist on any particular proportion of the capital expenditure of a scheme being devoted to art or craft' (Arts Council, 1990: 5,7).

15 'Comments such as "splendid atrium" were typical. One managing director ... was enthusiastic about the facility provided by the atrium in his building ... provided a better space for problem-solving discussions than the conventional meeting-room' (Roberts *et al.*, 1993: 20).

16 The factors with ratings (out of 10) are: location/locality 8; rental cost 8.4; quality and type of office 8.3; useful/compatible occupiers 3.4; image/attractiveness 6.8; public art 4.5. (Roberts *et al.*, 1993: 22) It could be argued that art contributes to attractive-ness, but presumably not as much as design.

17 'The problem is that the loss of democratic accountability in these ways is less dramatic, and less visible to the electorate, than would be other means by which a similar diminution of democratic power might come about' (Ambrose, 1994: 211). Rosler writes of 'the current high-profile version of "beautification", an ambition to improve the "quality of life" often invoked by anxious city administrations in cancelling both taxes and unsightly urban elements for the benefit of powerful corporations' in relation to Battery Park City (Rosler, 1991: 31).

18 Waterfronts provide not only disused dock areas (following new technologies of bulk transport and the demise of ship building), but also the vista, the view which distances. For comparison of waterfront developments, see Proudfoot, 1996; and *Urban Design*, no. 55, July 1995: 18–36.

19 The 1993 Arts Council/British Gas Award for Working for Cities.

20 Whyte (1980) is cited in the *Strategy*'s booklist, but has not informed the discussion within it.

21 The benefits listed are: tourism, cultural profile, advertising, product awareness, contact with opinion formers, entertainment opportunities and company prestige.

22 This was recommended by Michael Diamond, Director of Birmingham City Museums and Art Gallery, as early as 1982, in the initial planning stage and well before the Arts Council campaign. The policy produced a budget of £740,000; due to escalating costs, this was finally 0.44 per cent of the capital scheme. Sponsorship by the Public Art Commissions Agency raised £200,000.

23 Hasleden is reported as seeing his sculpture as offsetting the bleakness of the building, a view which might not be shared by the architects, Percy Thomas Partnership.

24 Gormley's piece was commissioned by the TSB and stands near their offices. At the time of writing there is talk of its being moved to Bristol, which raises an interesting question as to whether the work is 'site-specific' or 'site-general'.

25 The original proposal for Victoria Square was by Marta Pan, rejected as a result, according to one senior Council officer, of rivalry between departments. But, at Marta Pan's presentation of her proposal to a conference in Birmingham in 1990 in terms of 'The change in level ... offers the possibility to create a composition of steps ...',

David Reason responded that 'It should be unacceptable for anyone dealing with public spaces to say that a difference in levels gives the opportunity to do something with steps. Public spaces are not spaces of consensus and community, they are spaces of confrontation and struggle and for the establishment become spaces of legitimate exclusion'. The outcome of the dispute within the local authority was that an Arts Council officer was brought in as a consultant (setting aside his post for the duration) to mediate. The square reads as a picturesque tableau from the lower level, but from above its elements seem crowded, like a Mannerist painting.

26 See Featherstone, 1995: 102–13; and Massey, 1994: 117–24. Featherstone questions the models of anthropology which do not give enough weight to the connection of even isolated rural settlements to global communications, and the nostalgia of coherences projected onto past models of settlement.

27 Fleming and von Tscharner write of the threat of standardised urban architecture linked to anomie, and propose that: 'works of public art, urban design artefacts . . . can help define, reveal, enrich, expand or otherwise make accessible the meanings of a particular environment' (Fleming and von Tscharner, 1987: 2). This suggests, in terms such as 'reveal' an approach which privileges received rather than socially produced meanings of site.

28 The correspondence between the numbers denoting space and expenditure on art are coincidental, the latter taken from a press release dated 5 December 1991. The following works are in outdoor spaces: *Fulcrum* by Serra; *Rush Hour* by Segal; *Bellerophon Taming Pegasus* by Lipchitz; *Hare* by Flanagan; *Ganapathi and Davi* by Cox; *Venus* by Botero; a tiled fountain by Artigas; gates by Evans; and *The Broad Family* by Corbero. Hodgkin's mosaic is in the health club; other works are in foyers (where photography is forbidden) and offices. There is a more recent sculpture by Bruce McLean in Bishopsgate, at the edge of the development.

29 Of nineteen pieces in place by January 1991, only one was by a woman artist (Therese Oulton).

30 The fact that the site is landfill has been taken to put it 'outside history'; but Deutsche writes: 'Of course, Battery Park City does have a history . . . a sequence of conflicts over the use of public land and especially over the composition of city housing' (Deutsche, 1991a: 191). The initial plan to include 1,400 units of low-rent housing was dropped from the development – for the history of this see Deutsche, 1991a: 192–4.

31 This could be compared with Zukin's formulation of an approach to critical thought about the city as a 'symbolic economy' which: 'focuses on representations of social groups and visual means of excluding or including them in public and private spaces' (Zukin, 1996: 43).

32 See Holub, 1992: 6 and references to hegemony in Chapter 3. Deutsche does not cite Gramsci, but his formulation is useful in this context.

33 'St Peter's is not unusual in being one geographic community containing many communities of interest. There is a community of long established families who have lived through the radical transformations of St Peter's. There is a community of more recent arrivals who have moved into the new housing estate. There is the old primary

school community and the new; there is the religious community and the secular; a community of those employed in traditional industries and another of those employed in new industries; there is a community with no job at all' (Artists Agency, 1996: 9).

34 The writer visited the site in July 1996 to take photographs and was constantly accosted by local people giving positive responses to the *Red House* and other works. This is an unusual experience; more often, elsewhere, local people have been cynical.

35 References to Pym 1995 are to the evaluation report drafted by artist William Pym in June 1995.

36 Selwood, using focus groups and interviews, also relied on anecdotal rather than scientific evidence. It is only in specific situations, such as the recovery of patients from illness, that controlled studies have been conducted to find the effects of different environmental factors.

6 ART AND METROPOLITAN PUBLIC TRANSPORT

1 Art in transport has its own conferences, such as Junction 96 in Lisbon (28 September– 1 October, 1996), and was the subject of a report, 'Travelling Hopefully' for the Gulbenkian Foundation (UK) in 1992 – see Khan and Worpole, 1992.

2 Cases include the design of new metalwork in the New York Subway, new bus design in Seattle, and station design in St Louis, as well as London Underground's station upgrading of the 1980s and the Stockholm Metro.

3 For example, *Nine Spaces, Nine Trees* by Robert Irwin, 1983.

4 See Girardet, 1992: 148, and Richards, 1990. Richards gives the examples of Hamburg and Hanover as having progressive public transport policies. One of the most successful systems is the metro bus network in Curitiba, Brazil. At the same time, metropolitan transport contributes to suburbanization and will be less necessary if mixed zoning enables more people to live, work and shop in the same small areas.

5 Seeing the relation of art and transport from another perspective, public transport creates access to cultural provision, usually situated in town centres. See 'Creating Choices', a strategy document produced by West Midlands Arts in 1993, pp. 49–50.

6 This still raises a question as to which publics are addressed by a systems public image.

7 See *The Guardian*, 15 July 1996, p. 7.

8 See in particular Raven, 1993: 139–54 for a chapter by Wendy Feuer on the New York Subway; and Abramson, 1995 for a critical review of art in metro systems; also Petherbridge, 1989: 56–8 and Plates 10 and 12 for a description by Lesley Greene on art in European metros; More, 1984 on the Buffalo Underground; Shamash, 1991: 36–9 on Seattle Metro Bus Tunnel; Khan and Worpole, 1992: 35–6 on the Stockholm Metro. For a discussion of the durability of public art with reference to cases in metro stations, see an article by Marianne Strom in *Environnemental*, 1993 vol. 6–9, 304–5.

9 Amongst the more ambitious examples of airport art are Michael Hayden's neon *Sky's the Limit* at Chicago's United Airlines terminal, which uses 466 sequentially programmed neon tubes in a spectral range of colours, and William Pye's stainless steel cone fountain at Gatwick's North Terminal. For a description of art in Australian

airports, see Battersby, 1994. Hayden produced a neon piece also for Yorkdale Station, Toronto.

10 See Chapter 2.

11 'Guimard's shelters have become synonymous with the Metro and with the city of Paris itself and serve as local landmarks in the neighbourhoods where they still stand' (Abramson, 1995: 74).

12 'The state of its transport system can be seen as a sign of a country's values. In the old days, Moscow's metro had a magic that seemed to signify a government that believed the common citizen merited a system that looked – with its marble and chandeliers – like a palace' (Khan and Worpole, 1992: 1.11).

13 'August P. Belmont, the banker who financed the Interborough Rapid Transit Company (IRT) – the first subway line – was said to have been especially proud of the decorative elements in the stations, for which he had budgeted $500,000' (Feuer, 1993: 140). 'When work began on the first New York City subway stations in 1900, construction contracts stipulated that each station be embellished with decorative elements to "uplift the spirits of riders". This standard was regularly met with striking examples of mosaic tile work, marble wainscoting, terra cotta bas-reliefs with associations to communities above, and cast iron station entrances' (MTA, n.d.).

14 Alan Kiepper, President of the New York City Transit Authority, recalls in particular the art in San Francisco's BART and the music in Brussels – 'These were efficient public transit systems. But they also had a personality that reflected and enhanced the civic character of their cities. They were public systems that embraced public art in the fight to attract riders from their dependency on cars' (Kiepper, 1994: 15).

15 Wendy Feuer, Director of Arts in Transit writes: 'These contemporary symbols of travel, local scenery, and other objects all done in a hieroglyphic motif, provide subtle directional signals as they make a playful reference to place. They are, in a sense, an update of the subway's original faience plaques; on one level telling people where they are, on another telling them where they are going' (Feuer, 1993: 142–3).

16 'The Underground traditionally enjoyed a world-wide reputation as a patron of the arts. This reached its peak under the enlightened management of Frank Pick ... In the station work of the 1980s London Underground Limited sought to re-establish that tradition as part of a new design strategy' (LUL, 1993: 7).

17 'Holborn Station open competition for artists', LT Executive, department of the Chief Architect. The competition was supported by the Arts Council of Great Britain; the closing date for entries was 14 March 1980, and the first prizes (of which five were offered) £1,000.

18 'When viewing the site, all London Transport Byelaws are to be observed and competitors will not be admitted to the site without a valid ticket. As Holborn station is one of the busiest on the Underground, competitors will be expected not to hinder the flow of passengers in any way. It is suggested that the site be viewed out of the rush hours. Please notify station staff if you wish to take photographs. Flash photography is not permitted' (Holborn Station: open competition for artists).

19 Pick (1878–1941) was qualified as a solicitor and experienced as a statistician, but

joined London Underground following the appointment of Sir George Gibb as Managing Director in 1906. He is quoted as saying: 'After many fumbling experiments I arrived at some notion of what poster advertising ought to be. Everyone seemed to be quite pleased with what I did and I got a reputation that really sprang out of nothing' (Green, 1990: 9). He was the first Chair in 1934 of the Council for Art in Industry, forerunner of the Design Council.

20 A poster designed by Alfred France in 1911, illustrated in Green, 1990: 23. At the same time, some posters appeal to a specifically middle-class audience, for example two designs with the text 'Brightest London' by Horace Taylor in 1924; one depicts diners in evening dress, the other a similar social group on the escalators – illustrated in Green, 1990: 70–1.

21 Illustrated in Green, 1990: 72.

22 Since the policy was revived in 1986, six new images a year have been commissioned, some by artists of national standing, including Howard Hodgkin and John Bellamy. Of the six, two are widely accessible, two 'mainstream' and two avant-garde. Six thousand copies are printed of each and are both pasted up and sold to the public.

23 Illustrated in Green, 1990: 135.

24 Charing Cross, Baker Street, Marble Arch, Finsbury Park, Paddington, Holborn, South Kensington, Waterloo, Mansion House.

25 Tottenham Court Road, Euston, Embankment, King's Cross.

26 Heathrow Terminal 4, Liverpool Street, Angel, Hammersmith, Monument.

27 Note supplied to the author by the artist, 1993.

28 See discussion in Chapter 3.

29 Waterloo International was RIBA Building of the Year in 1995.

30 Amongst those depicted in Ringgold's design and others by Willie Birch and Vincent Smith are Adam Clayton Powell Jr, Langston Hughes, Duke Ellington, Billie Holiday and Malcolm X (MTA, 1994: 9).

31 The advocacy of public art in the UK generally emphasised the role of professionals, and marginalised community arts. It has given little attention to the cultural identities of black communities.

32 'Commissioned artworks became the norm after Seattle's 1% for Art ordinance (105389) was enacted . . . The Seattle City Council recognized that the "development of the visual and aesthetic environment is a fundamental responsibility of civic government"' (Rupp, 1991: 15)

33 For further description see Shamash, 1991: 36–8.

34 'Before long, though, the design team found itself being pulled in different directions by the Arts Committee, which had hired the team and met with it monthly to discuss aesthetic issues, and the employee task force, which met almost daily with the team and provided technical information'; the conflict was resolved through cross-representation (FTA, 1996b: 27).

35 A value-engineering process (sometimes called value management elsewhere) was used to manage costs, through which some artist-designed elements were rejected (FTA, 1996b: 25).

7 ART IN HEALTH SERVICES

1 See Rubik, 1992

2 See Foucault, 1967, on the establishment of the General Hospital in Paris; for a history specific to the institutionalisation of insanity in nineteenth-century England, see Walton, 1985. Walton states. 'A further important aspect of the medical model of insanity was that a widening range of problematic behaviour and circumstances could be blamed on the medical problems of individuals rather than the deficiencies of society at large' (p. 133).

3 'In the medical tradition of the eighteenth century, the disease was observed in terms of *symptoms* and *signs*. These were distinguished from one another as much by their semantic value as by their morphology. The symptom . . . is the form in which the disease is presented: all of that is visible . . . The sign announces: the prognostic sign, what will happen; the anamnestic sign, what has happened; the diagnostic sign, what is now taking place' (Foucault, 1973: 90).

4 Referencing the *Dance of the Dead* in the Cimetière des Innocents in Paris (now lost but known through copies), Illich writes: 'In the shape of his body Everyman carries his own death with him and dances with it through his life' (Illich, 1976: 177).

5 'The new class of old men saw in death the absolute price for absolute economic value. The ageing accountant wanted a doctor who would drive away death' (Illich, 1976: 193).

6 The difference is epitomised by the contrasting projects in the Manchester hospitals, where a team of artists worked fairly like a community art group, and St Charles Hospital, London, where a Beautification Committee chaired by a Consultant Physician oversaw a project which included both restoration of the buildings and work by contemporary fine artists – See Baron, Hadden and Palley, 1984.

7 See Breslow, 1982.

8 See Greene, 1987: 19; decorative tiling by Minton is found in the entrance hall, the Board room, the staircase and corridor of the Nurses' Home, and externally on the main entrance. Burmantoft faience is found in the adjacent Medical School of 1894.

9 Another reading of the 'Nightingale' ward is in terms of Foucault's discussion of the panopticon in *Discipline and Punish*, in which observation becomes surveillance.

10 The scale of the Victorian asylum represented a massive increase in the medicalisation of behaviour found problematic by the guardians of society's morality – pregnancy in the servant class, for example, being regarded as a sign of moral deficiency. For a history of a more human-scale asylum in which security was combined with architectural grace, the York Retreat opened in 1796, see Digby, 1985. Digby cites the Quaker William Tuke (of the York Retreat) as wishing to avoid the 'gloomy appearance [of] places appropriated for those who are afflicted with disorders of the mind' (p. 54), and notes (but questions) Foucault's critique of such 'humane' institutions as disciplining through a system of rewards and punishments for conformity.

11 St Nicholas Hospital, Gosforth, Newcastle, has a recreation hall stage with a proscenium arch of Doulton tiles in an Art Nouveau design (1900) by W. J. Neatby, who designed the tiles for the food hall in Harrods (Greene, 1987: 50). Walton draws

attention to the familiar complaint that lunatics enjoyed 'luxurious conditions . . . at extra cost to the rate payers', a level of (non-luxurious by most standards) care protected by the appointment by central government of justices of the peace to the Boards of county asylums (Walton, 1985: 133).

12 See Greene, 1987. Greene gives a list of the benefactors providing the cost of the sixty tile pictures at the Royal Victoria Infirmary (the building itself funded by public subscription between 1896 and 1906) as the Misses Stephenson (probably daughters of the Lord Mayor), Mrs Ward and friends, the clinicians, and the Workmen Governors (p. 47).

13 A conference report on Art in Hospitals was published by the Department of Health & Social Security (DHSS) in 1983 (DHSS, 1983); guidelines for art projects in new hospital buildings were published by the DHSS in 1985 (Coles, 1985); Health Building Note 1, revised in 1988, contained advice to include landscaping and art in hospital design (DHSS, 1988: 2.19); a report circulated to all NHS hospitals in 1993 includes the following policy: 'A decision to include art in a capital scheme should be made at the outset, and budget and source(s) of funding identified' (NHSE, 1993: 15).

14 See Palmer and Nash, 1993.

15 Ward used the standard staining and laminating processes of formica production, receiving technical support from the manufacturers; illustrated in NHSE, 1993: 16.

16 The design features old and new Dunfermline, and was sponsored by the Carnegie Dunfermline Trust; see Health Care Arts, 1994.

17 See NHSE, 1994a: 16, 18, 36.

18 See South Tees Health, n.d. – one of the first formulations of a policy in the NHS for 'putting people first'.

19 Cases include the colour-zoned Cossacks in Stafford General Hospital, and the matching of images to the colours of zones at St George's, Tooting; the latter is derived from a model developed by Annette Ridenour in San Diego, matching colours to familiar images to facilitate ease of memory.

20 This became considerably more important after the inception of NHS Trust (locally managed) status, from around 1990.

21 NHSE, 1993 states: 'Art can create a "cultural bridge" between the hospital and its locality by: a. being made by local professionals; b. being made in local schools; c. depicting the locality; d. relating to local traditions or industries; e. relating to different groups of people in the community' (NHSE, 1993: 15).

22 Acting through NHS Estates, its department for advice on all matters of hospital design, building and maintenance.

23 Projects took place in 1989–90 (outpatient departments); 1990–91 (outpatient and emergency departments); and 1991–2 (emergency and day treatment departments). Hospitals were invited to bid for funding, and selected on the basis of the quality of the scheme proposed, its viability, and the extent to which it represented a typical situation from which other staff teams could learn. Amongst those selected were: Queen Elizabeth Hospital, Gateshead; St George's Hospital, Tooting; The City Hospital, Nottingham; The General Infirmary at Leeds; Preston Royal Hospital; Hull Royal Infirmary; Southend Hospital; Peterborough District Hospital; Coventry and

Warwickshire Hospital; Hillingdon Hospital; Kent and Canterbury Hospital; Addenbrooks Hospital, Cambridge; Cumberland Infirmary, Carlisle; Frenchay Hospital, Bristol; Manor Hospital, Walsall; and Whipps Cross Hospital, London.

24 See NHSE, 1991; NHSE, 1992; NHSE, 1994b.

25 The writer was employed as a consultant by NHS Estates to provide this advice to the local project teams, which normally included about six to ten staff from users groups, at different levels of seniority and representing doctors, nurses and managers, plus the designer, who might be either in-house (from the district estates office) or external.

26 For feedback on some of these commissions, see Duffin, 1992; for description of some projects see Miles, 1992.

27 A significant step was taken by introducing a low, but wide, desk in the accident and emergency department at Southend, where it has been found that the level of violent incident has not increased and verbal abuse has decreased (NHSE, 1992: 32–5).

28 See NHSE, 1991; NHSE, 1992; NHSE, 1993; and NHSE, 1994b for staff and patient comments included in official reports.

29 Nick Ross reports that, according to a report produced by Oxford University, of 276 standard procedures in obstetrics, '88 have unknown effects and 60 can positively be shown to have no effect and should be abandoned' (Ross, 1994: 46).

30 'In medicine the testing procedure is inevitable. Any treatment has to be investigated. One cannot apply art in hospitals without examining the results. If we neglect research we will not appear credible' (Granaat, 1982: 103).

31 'Art is rebellious against the logical reason of commonness. Art is the most passionate revolution of man against his daily fate' (Granaat, 1982: 101).

32 Unpublished research carried out in partnership with Gloucestershire Royal Hospital NHS Trust, 1993–4. A higher pulse rate was noted in the refurbished rooms, explained by the soporific impact of dreary sites; readings for stress were based on patients' self-assessment.

33 The average stay for the 'tree' group was 7.96 days, compared with 8.7 days for patients whose only view was of a brick wall.

34 Strong doses during days 2 to 5 were 0.96 for the tree group, and 2.48 for the wall group.

35 See Heerwagen, 1990, for a summary and bibliography re windows; see also Gold, 1977 for general support for the social benefits of trees; and Relf, 1990 on horticulture and well-being.

36 See Dobbs, 1990, which observed the impact of four patterns of wallcovering – one warm with a vibrating pattern, one cool with the same pattern, one warm with a representational pattern, and one cool with the same pattern – and found 'no conclusive evidence of an identifiable relationship between selected wallcovering patterns and behaviour of residents. Review of the literature and anecdotal reports indicate that some patterns have been found to contribute to confusion . . . The findings of this study do not appear to support these results'.

37 'A work of art can function indirectly by improving the working climate for staff . . . Art can help in humanising relationships between staff and patients' (Granaat, 1982: 105).

38 Unpublished study conducted by arts management trainees supervised by Healthcare Arts, Dundee. See Miles, 1994, for a summary.

39 Peloquin, 1991: 'Art invites helpers to grow self-aware . . . to feel the experience of another . . . to take on the perspectives of others, to try out new meanings, and to step into horrible circumstances from a safe place'.

40 Unpublished responses to a proposal for patient-centred design at South Cleveland Hospital, Teeside, 1993; cf. Winkel and Holohan, 1985 cited above.

41 See Bertman, 1991.

42 Hospital routine and uniforms may be seen as elements in the distancing of staff from patients, as in the establishment of a chain of command on a military model.

43 'Beneath a gaze that is sensitive to difference, simultaneity or succession, and frequency, the symptom therefore becomes a sign . . . Analysis and the clinical gaze . . . compose and decompose only in order to reveal an ordering that is the natural order itself . . . for a doctor whose skills would be carried "to the highest degree of perfection, all symptoms would become signs", all pathological manifestations would speak a clear, ordered language' (Foucault, 1973: 94–5).

44 Therese Schroeder-Sheker writes on the consolation of death in Cluniac monasteries: 'In studying the Cluniac infirmary methods with the aid of anthropology, death is seen as a *rite of passage* composed of a three-part structure: the rites of separation, the rites of *liminality* . . . and the rites of reincorporation . . . Readers should not confuse this structure with other psychological models of grief and loss work' (Schroeder-Sheker, 1993: 43).

45 'Only a culture that evolved in highly industrialized societies could possibly have called forth the commercialization of the death-image that I have just described' (Illich, [1984] 1986: 206) and 'Dying has become the ultimate form of consumer resistance' (ibid.: 207).

46 'TB is described in images that sum up the negative behaviour of nineteenth-century homo economicus: consumption; wasting; squandering of vitality. Advanced capitalism requires expansion, speculation, the creation of new needs . . . Cancer is described in images that sum up the negative behaviour of twentieth-century homo economicus: abnormal growth; repression of energy, that is refusal to consume or spend' (Sontag, 1990: 63).

47 See NHSE, 1994a: 16.

48 Originated at the Lakeland Medical Centre, Florida in 1988.

49 The model is similar to that used for some years by progressive industry, the Volvo plant in Sweden, for example, operating through teams who build whole cars rather than a production line.

50 See Karanian, 1994.

51 See Filochowski and Cadenhead, 1995.

8 ART AS A SOCIAL PROCESS

1 The art market has tended to absorb and turn into profit almost all anti-art movements to date, including that of the 1960s; if that does not happen with new genre public

art, a reason may be the higher degree to which the artists are politicised and prepared to take responsibility for their own futures.

2 Its politicisation includes the feminist position that 'the personal is political'.

3 Gablik uses remarks by Barbara Rose, from an article in *Vogue*, as a foil to her own position, summarising Rose: 'This willingness to risk confrontation with the audience . . . is precisely what gives modern art its moral dimension . . . Thus, for Rose at least, Serra's work, with its almost mineral imperviousness, is the ultimate model of social independence and the radically separative self' (Gablik, 1991: 63). Gablik utilises the 'dominator' model of patriarchy from Riane Eisler's *The Chalice and the Blade* (in contrast to the 'partnership' model) as a framework for Serra's work; 'Within the "dominator" system, the self is central; power is associated with authority, mastery, invulnerability and a strong affirmation of ego-boundaries' (ibid.: 62).

4 See Tilyard, 1963, for references to a 'centre which does not hold' in Elizabethan literature.

5 'Because . . . modernity has failed to fulfil its promises of "a better life" in many of the deepest senses, we are compelled to search for new, or perhaps recovered, modes of understanding our nature and the relation between our species and the rest of the natural world' (Spretnak, [1991] 1993: 12) 'The remythologizing of consciousness, then, is not a regressive plunge into the premodern world . . . it represents a change in how the modern self perceives who it truly is, when it stretches back and contacts much vaster realities than the present-day consumer system of our addicted industrial societies' (Gablik, 1991: 57).

6 'A belief that resistance to the dominant social structure is futile, because the structure is too ruthless or powerful, will have the effect . . . of stabilizing the relations of dominance' (Gablik, 1991: 23).

7 This does not include art therapy, which is a clinical practice within, if marginally and uneasily so, the medical establishment.

8 The categories of potential new member listed by the Art Studio in Sunderland include people discharged into the community under the care of Sunderland Health Commission, people suffering phobias, people with learning disabilities, people with serious loss of confidence through long-term illness, people suffering from schizophrenia, people suffering bereavement, anxiety or depression, or substance abuse, and primary care service clients.

9 In 1996, twenty members had moved to employment, six to further education and five to full-time art practice outside the Art Studio.

10 The Art Studio in South Shields operates in partnership with South Tyneside Metropolitan Council and South of Tyne Health Commission.

11 A health visitor outlined specific mental health problems due to discrimination and harassment, communication barriers, poverty and immigration laws, as well as generational conflict.

12 See Frohnmayer, 1993: 132–4 and 142 for an account of attacks on arts funding in the USA following the exhibition of photographs by Robert Mapplethorpe in Cincinnati in 1991.

13 Watney takes issue with the 'anti-monuments' of artists such as Jochen and Esther

Shalev-Gerz, on the grounds that invisible monuments equate with invisible crimes against humanity which art should make visible: 'The last thing we need . . . are the pretentiously over-intellectualised justifications for politically correct "invisible monuments"' (Watney, 1994: 60).

14   The smaller text states: 'Corporate Greed, Government Inaction and Public Indifference Make AIDS a Political Crisis' (Meyer, 1995: 51); Gran Fury was an anonymous group of about fifteen members of ACT-UP, including artists, a hairdresser, a costume designer and an architect, formed following the project *Let the Record Show* (Meyer, 1995: 64). The posters are also cited by Ted Gott in *Don't Leave Me This Way*, a publication following an exhibition of art in the age of AIDS at the National Gallery of Australia.

15   Illustrated in Meyer, 1995: 55; the text continues 'Unnaturally grown with insecticides, miticides, herbicides, fungicides'.

16   For an account of this censorship see Engberg, 1991.

17   Press coverage is increasingly seen as part of an art strategy – see also the description of the *Welcome to America's Finest* posters under Cultural Diversity below. Gott cites Paul Taylor: 'Even better for Gran Fury, many of the papers quoted the text on the billboard, thereby spreading its message about Safer Sex way beyond the relatively elite crowd that visits the Biennale' (Taylor, 1992: 29).

18   Illustrated in Meyer, 1995: 83.

19   The 1988 poster includes a photographic image, taken by Sisco on a San Diego bus, of an immigrant handcuffed to a border agent, flanked by images of a dishwasher and chambermaid, two areas of service dependent on immigrant, low-paid labour. Efforts to suppress the poster, for which the artists rented space on buses, produced national press coverage. See Pinkus, 1995.

20   An earlier piece, *Cornered*, included a video monitor flanked by two birth certificates, one stating a black, the other a white, father; the artist began by stating 'I'm black. Now let's deal with this social fact and the fact of my stating it together'; sixteen monitors were used in the Whitney piece, each with a single white person speaking in a different language, around the reconstituted *Cornered*. Part way through, the sixteen peripheral voices suddenly turn to English to say 'Some of my female ancestors were so-called "house-niggers" who were raped by their white slave masters. If you're an American, some of yours probably were, too' (Lacy, 1995: 267).

21   For information on the controversy surrounding this project, see Doss, 1995: 1–11.

22   The posters and the development ('Dystopia on the Thames') are described in *Mapping the Futures* (Bird *et al.*, 1993: 120–49) and the posters in *Mapping the Terrain* (Lacy, 1995: 253–4).

23   Wehn produced a full-length bronze of Chief Seattle (1912) sited in Tilikum Place, see Rupp, 1991: 104.

24   Rupp writes: 'Totem poles are a three-dimensional art form generally associated with the Tlingit of southeastern Alaska, the Haida of the Queen Charlotte Islands, and the Tsimshian and Kwakiutl of western British Columbia. No Indian (sic) tribes in the Seattle area practiced this advanced and detailed art' (Rupp, 1991: 21).

25 For an account of the conflict see N. Coleman and J. Camp, 1988, *The Great Dakota Conflict*, St Paul, Pioneer Press. Billings wrote in 1990 that the work awaited reinstallation at the Land of Memories park in Mankato.

26 Jacobs draws attention to this as a frequently used strategy, whereby native title is re-conferred if land is then given over to national park use (Jacobs, 1996: 122).

27 The piece was commissioned by the developers of a shopping centre on a green site, and could be seen as a form of 'conscience money'. The artist has since attempted to find a new site.

28 These were commissioned through Common Ground and intended to encourage the use of footpaths by being wayside markers. The sculptures are not labelled in any way. Walkers have sometimes placed wild flowers in the niches.

29 Several extracts of Mazeaud's 'riveries' are cited by Gablik – Gablik, 1991: 119–21; the entry for July 20 [1988?] reads 'Two more huge bags I could hardly carry to the cans. I don't count anymore . . . I don't announce my "art for the earth" in the papers either . . . All alone in the river, I pray and pick up, pick up and pray . . . I feel the pain quietly' (Gablik, 1991: 121).

30 See Brigham, 1993.

31 The Wandle, Fleet, Effra and Walbrook.

32 Pig's Eye landfill, St Paul, Minnesota.

33 Plants which have a particular ability to take up toxins from the soil and store them in leaves and fruit.

34 See discussion of space and gender in Chapter 2.

35 See Crawford, 1992.

9 CONVIVIAL CITIES

1 'The level of urbanization is projected to cross the 50 per cent mark in 2005. UN projections further show that by 2025, more than three-fifths of the world's population will live in urban areas. The urban population in that year will be approximately 5.2 billion, of whom 77 per cent will live in developing countries' (Badshah, 1996: 2). This rate of growth, based on present trends of urban expansion, is certain to produce water shortages and deluges of waste, chaotic provision of housing and transport, and levels of abjection in response to which art will be powerless.

2 See Savitch, 1988.

3 See Gablik, 1991: 27.

4 Chantal Mouffe, 'Deconstruction, Pragmatism and Democracy', in Mouffe, 1996: 1–12. Mouffe concludes, 'In order to impede the closure of the democratic space, it is vital to abandon any reference to the possibility of a consensus that, because it would be grounded on justice or on rationality, could not be destabilized'.

5 Partners for Liveable Places, since renamed Partners for Liveable Communities, is based in Washington, DC; it published *Art in Public Places* (NEA, 1981) and currently produces a newsletter, *Livability*. It defines itself as 'a nonprofit organisation working to improve the livability of communities by promoting quality of life, economic development, and social equity' (*Livability*, October 1995).

6 Jacobs and Appleyard formulate their manifesto as an alternative to Le Corbusier's principles enshrined in the Charter of Athens of the International Congress of Modern Architecture.

7 'Where plural planning is practiced, advocacy becomes the means of professional support for competing claims about how the community should develop. Pluralism in support of political contention describes the process; advocacy describes the role performed by the professional in the process' (Davidoff, 1965, in LeGates and Stout, 1996, 425).

8 Alexander contrasts the contained sociation of traditional, or village society with those of industrial urbanism: 'In a traditional society, if we ask a man to name his best friends and then ask each of these to name their best friends, they will all name each other' (Alexander, 1965, in LeGates and Stout, 1996: 123); this may be a romantic view of village society, found in the work of the Chicago School in the 1930s, for whom the rural settlement tended to be defined as a foil, separated in time, to the urban settlements they studied. Alexander goes on to argue that planned cities take on closed attributes, and that only cumulative development and appropriation lead to the semi-lattice of a healthy city.

9 Sennett writes of the playgrounds of Battery Park City: 'The few infants cavorting are happy enough; older children seem at a loss for what to do' and of SoHo: 'It is a community surprisingly dense with children who have a healthy ragamuffin energy. They have colonized the loading docks, the empty lots, the roofs . . .' concluding 'once they can go out alone, these children avoid the "designer" playground . . . ' (Sennett, [1990] 1992: 193–4).

10 Sennett explains disengagement through the separation of inner and outer life in the Protestant ethic of inwardness at the expense of (splitting from) engagement with outer realities: 'There is withdrawal and fear of exposure, as though all differences are potentially as explosive as those between a drug dealer and an ordinary citizen' (Sennett, [1990] 1992: 129).

11 See Catchpole, 1996, on Agenda 21.

12 'The message from the tenants (both from the study and our earlier meetings) was clear: They wanted to live in a "normal" neighbourhood, one that didn't look or work like a project, one that felt safe for walking around and letting their children out to play' (Goody, 1993: 23).

13 See 'Winners in the OXO game' by John Cunningham, *The Guardian*, 11 Sept. 1996.

14 There is also a danger that such solutions will romanticise the idea of urban settlement using a model derived from (possibly false) ideas of the rural village, whilst some urban dwellers, particularly single people (who constitute a growing proportion of the population) may like to live on the upper floors of high-rise buildings, enjoying the view and seclusion.

15 A comparison could be made with the utopian experiments of the nineteenth century, such as the Shakers (1787–present), Harmony Society (1804–1904) and Oneida (1848–81), all in the USA. McLaughlin and Davidson report that, between 1825 and the American Civil War, more than a hundred new communities were established, many influenced by the socialism of Robert Owen and Charles Fourier (McLaughlin and Davidson, 1985: 87).

16 Landry and Bianchini also argue for cultural provision as part of 'the creative city' – see Landry and Bianchini, 1995: 47.

17 Miffa Salter, with Rachel Dunlop, 'Seeking Quality Assurance in Urban Design', forthcoming article in *Built Environment*.

18 'It was the transport and communications revolution of the nineteenth century that finally consolidated the triumph of space as a concrete abstraction ... land became nothing more than a particular kind of financial asset' (Harvey, 1989: 177).

19 'History presents us with a wide range of unique and distinctive cities ... the more dynamic Greek democracies [centring] around civic squares that fostered citizen interaction' (Bookchin, 1992: 5); Bookchin aligns the polis with the absence of a state in the modern sense: 'when the Athenian democracy was approaching its high point of development, the concept of a state ... was notable for its absence' (ibid.: 33).

20 'The designer sees the whole building ... the person sitting on the plaza may be quite unaware of such matters' (Whyte, 1980: 26) and 'Urban designers believed [shape] was extremely important ... Our data did not support such criteria' (ibid.).

21 For relation of Greenacre Park to the introduction of conservation easements as charitable tax deductions in 1986, see Whyte, 1988: 281.

22 The challenge is taken up by artists who have produced designs for gardens in New York, including Vito Acconci, Alison and Betye Saar, Gary Symons and Meg Webster; their proposals are detailed in *Urban Paradise: Gardens in the City*, published by Public Art Fund in 1994.

23 Salter continues: 'Thus consultation also has a role as a conduit between various stakeholders' (Salter, op. cit., – see note 17).

24 Examples include a downtown plan for San Bernadino, California, projects for urban transport facilities in Corpus Christi, Texas and New Jersey, an open spaces programme for Hartford, Connecticut, and the redevelopment of Block 57 in Salt Lake City, Utah. At Salt Lake, PPS 'worked closely with community groups in workshops to prepare a program of uses and an overall plan in order to ensure that the square becomes a major activity center, accommodating uses attractive to downtown employees ...' (PPS publicity material, n.d.).

25 See Spitzer and Baum, 1995, on public markets and community revitalisation.

26 A parallel is the empowerment of patients in the Planetree model of community health care, discussed in Chapter 7.

27 Illich writes: 'The demands made by tools on people become increasingly costly. The rising cost of fitting man (sic) to the service of his (sic) tools is reflected in the ongoing shift from goods to services ... Increasing manipulation of man (sic) becomes necessary to overcome the resistance of his vital equilibrium to the dynamic of growing industries' (Illich, 1973b: 46).

28 'Language reflects the monopoly of the industrial mode of production over perception and motivation. ... The functional shift from verb to noun highlights the corresponding impoverishment of the social imagination' (Illich, 1973b: 89). As an example Illich gives the statement 'having work' rather than 'working' or 'doing' work.

29 Cases of UDATs include those for Wood Green and Burgess Park in London; the latter comprised 130 acres of 'green space which nobody wanted' and over which discussions

were not easy – John Worthington writes 'it is arrogant to think you can come in for two and a half days to look into the problem . . . The presentation was a holocaust . . . It reflected twenty years of sheer frustration' (Worthington, 1996: 13); Worthington notes Davidoff's work as the beginning of this approach, followed by the establishment of a community architecture group at the RIBA in the early 1980s.

30 Rowland gives the example of a housing project in which car parking provision was based on an inappropriate model: 'the result is an urban desert. Similarly, open space is given as a ration of estimated population based on habitable rooms, not use. How the open space is utilised is not defined. The end product is lots of useless space around winding cul-de-sacs, or parking courts' (Rowland, 1995: 24).

31 Gare attributes the term 'polyphonic narrative' to Mikhail Bakhtin, and cites Julia Kristeva's position that such a narrative: 'constructs itself through a process of destructive genesis' (Gare, 1995: 140).

32 See Lavenda, 1992, for a critical account of festivals and the voices they represent.

33 From an unpublished paper supplied to the writer.

34 From a paper distributed at a conference in Manchester, 1994.

35 'It has been used for target practice by local air rifle enthusiasts and is currently fenced off prior to demolition. This expensive failure calls into question the wisdom of placing costly art objects in vulnerable, undefendable places and of confronting an unsympathetic public with the excesses of "Turner Prize" modernism' (Editorial, *Art Review*, July/August 1995). The project was managed by Public Art Development Trust, which the editorial states receives £120,000 annually in public funding.

36 From a 'curatorial statement of program need' as part of a grant application, 1995/6.

FURTHER READING

•

INTRODUCTION

As an introduction to the main subject areas of this book, *The City Reader* (LeGates and Stout, 1996) offers a wide range of texts, several otherwise difficult to obtain, organised in sections dealing with the evolution of cities, urban form and design, urban culture and society, urban politics, urban planning and the future of cities. Writers whose work is extracted include historical figures such as Friedrich Engels and Frederick Law Olmsted, those influential on the formation of the discipline of urban and regional planning, such as Patrick Geddes, Louis Wirth and Ernest Burgess, and more recent urban critics such as Jane Jacobs, Mike Davis, Dolores Hayden and Saskia Sassen.

There is no corresponding reader on art in public spaces, but Suzanne Lacy's *Mapping the Terrain* (1995) provides a critical introduction to the subject and a 'map' of current socially concerned practice.

1 THE CITY

Richard Sennett's *Flesh and Stone* (Sennett, 1994) constructs a narrative of city form as expression of social attitudes to the body, from Greece to New York. His earlier works, *The Uses of Disorder* (1996), *The Fall of Public Man* ([1974] 1992), and *The Conscience of the Eye* ([1990] 1992), constitute a linked set of considerations, mingling his experience of cities such as New York with history and theory on the alienation of the modern city.

Elizabeth Wilson's *The Sphinx and the City* (Wilson, 1991) takes personal memories as a point of departure for an analysis of the position of women in the cities of modernity and constructs a radical critique of urban development. For Wilson, the city can be celebrated as a location of freedom, though its potential faces destruction by 'developers', and the extension of freedom to diverse urban publics requires a new kind of planning.

Re-Presenting the City, edited by Anthony King (King, 1996) is a collection of papers originating from a symposium on ethnicity, capital and culture in the metropolitan cities of the twenty-first century; it includes contributions by geographers James Duncan and Neil Smith, sociologist Sharon Zukin, and planner Saskia Sassen. Unlike many works on the city, including this one, it pays considerable attention to cities in non-western countries, and brings together arguments which constitute a radical critique of the western, industrial metropolis.

Strangely Familiar, edited by Iain Borden, Joe Kerr, Alicia Pivaro and Jane Rendell (Borden *et al.*, 1996) includes brief texts and photo-essays, by writers from various academic disciplines, including Barry Curtis, Dolores Hayden, Doreen Massey, Edward Soja and Elizabeth Wilson, on the question 'what do our cities mean to us?', taking cases from London, Amsterdam, Venice and Sao Paulo, amongst others.

2 SPACE, REPRESENTATION AND GENDER

The key work on urban space is Henri Lefebvre's *The Production of Space*, first published in 1974 but in English in 1991 (Lefebvre, [1974] 1991). It is a difficult and wayward book, intertwining several areas of enquiry and retrospect, on French intellectual life, on Marxism, on urban space; but it formulates the model of representational spaces (of experience, the street) and representations of space (concepts, the plan) under the heading of spatial practice, and establishes that space is socially produced, not given.

Other works by Lefebvre in English translations include *Critique of Everyday Life* (Lefebvre, [1947] 1992) and *Writing on Cities* (Lefebvre, 1996) a collection of short, generally accessible pieces. Michel de Certeau's *Practice of Everyday Life* (de Certeau, [1984] 1988) articulates a similar territory.

A feminist framework is given, particularly in the second and third parts, in Doreen Massey's *Space, Place and Gender*; this represents new thinking in geography which re-introduces discussion of space and place, as well as giving personal experiences which illustrate a gendered public realm. Elizabeth Wilson's *The Sphinx in the City* is as relevant to this chapter as to the previous, and *Body Space*, edited by Nancy Duncan (Duncan, 1996) includes texts by cultural geographers which investigate the idea of knowledge embedded in place and space. Luce Irigaray's proposal to display images of the mother and daughter is contained in 'A Chance to Live', one of the essays in *Thinking the Difference* (Irigaray,1994).

A contribution, again from cultural geography, which relates to discussion of both cities and space in terms of the ways society produces abjection through the exclusions and boundaries within the city, is David Sibley's *Geographies of Exclusion* (Sibley, 1995). Sibley's arguments refer to issues of gender, race, sexuality, disability and age, and the links between them.

3 THE MONUMENT

Two critical works are useful: William Mitchell's collection of essays by various authors on monumental public art, *Art and the Public Sphere* (Mitchell, 1992), which covers both conventional memorials to wars and generals, and recent cases such as the *Vietnam Veterans Memorial*; and Samir al-Khalil's *The Monument* (al-Khalil, 1991), a detailed contextual and critical study of Saddam Husain's Victory Arch, which addresses the issues of kitsch, and the framework of heritage culture in establishing the inevitability of power.

Jon Bird's catalogue essay 'The Spectacle of Memory' for the Whitechapel Gallery exhibition of sculpture by Michael Sandle (Bird, 1988) includes a summary of the relation of memorials to national identity.

Jo Darke's *The Monument Guide* (Darke, 1991) is a descriptive text listing, and often illustrating photographically, hundreds of monuments in England and Wales. A descriptive catalogue of public sculpture, including many nineteenth-century examples, in New York is given in *Manhattan's Outdoor Sculpture* (Gayle and Cohen, 1988). Both volumes include useful material on which the authors' comments are incidentally interesting.

Public Address (Freshman, 1992) includes critical essays on Krzysztof Wodiczko's projection pieces by Dick Hebdige, Patricia Phillips, Andrzej Turowski and Peter Boswell, with statements by the artist and many illustrations.

4 THE CONTRADICTIONS OF PUBLIC ART

John Willett's *Art in a City* (Willett, 1967) remains interesting, and many of its questions remain unanswered by the public art establishment. It could be read in conjunction with *Patronage and Practice – sculpture on Merseyside* (Curtis, 1989), which gives several accounts of both Victorian and recent public sculpture, from the monument to Lord Nelson (1807–15) to the Liverpool Garden Festival (1984), and two essays, by Ronald Alley and Lewis Biggs, on the sculpture collection of the Tate in Liverpool.

Selwood's *The Benefits of Public Art* (Selwood, 1995), her report for the Policy Studies Institute, is the only work of its kind and should be required reading for all arts managers and potential patrons of public art. Its emphasis on mechanisms would be balanced by the more discursive writing and wider cultural references of *House* (Lingwood, 1996).

Amongst critical works, *Mapping the Terrain* (Lacy, 1995) and *But is it Art?* (Felshin, 1995) are in themselves maps of the emerging territory of activist or socially committed public art; these are both published by the Bay Press in Seattle, from whom a series of 'Discussions in Contemporary Culture' supported by the Dia Art Foundation offers a wider contextual framework. In this series are *Out of Site* (Ghirardo, 1991) and *If You Lived Here* (Wallis, 1991), both of which relate to public art and include important essays by Rosalyn Deutsche, and two volumes edited by Hal Foster: *Discussions in Contemporary Culture* (Foster, 1987) and *Vision and Visuality* (Foster, 1988).

A comparison of the issues of public art with those of curatorship can be made through Marcia Pointon's *Art Apart* (Pointon, 1994) and *Museums and Communities* edited by Ivan Karp, Christine Kreamer and Steven Lavine (Karp *et al.*, 1992). *Free Exchange* (Bourdieu and Haacke, 1995) includes a conversation between Bourdieu, known in particular for his idea of 'cultural capital', and Haacke, as well as cases of Haacke's projects.

5 ART IN URBAN DEVELOPMENT

Further reading on art in urban development is of two kinds: that found in critical approaches to public art, and that of the wider literature on urban development from a sociological viewpoint.

The texts cited above by Rosalyn Deutsche, in *If You Lived Here* (Wallis, 1991) and *Out of Site* (Ghirardo, 1991) are lengthy and closely-knit, their arguments requiring close reading and consideration of a kind not possible in the brief summaries given here. They offer a model of criticism which could be applied to many other developments.

Saskia Sassen's *The Global City* (Sassen, 1991) is a critique, contextual to this whole discussion, of corporate development and its establishment of glamorous and technically advanced enclaves within cities, which tend then to dominate the city in which they have been implanted.

The Living City by Roberta Gratz (Gratz, 1989) gives more descriptive accounts of a range of projects to reclaim and revitalize downtown districts in cities in the USA. Such projects, however, can be seen as alternatives to mega-corporate development and the profusion of out-of-town malls.

Urban Process and Power, by Peter Ambrose (Ambrose, 1994) deals specifically with the problems of development in the UK, in context of the neo-liberal ideology which has characterised schemes such as Docklands. It includes statistical information as well as argument and anecdote. Its connecting thread is the question of who controls urban development in a supposedly democratic society.

6 ART AND METROPOLITAN PUBLIC TRANSPORT

Transport in Cities (Richards, 1990) gives details, though little critical appraisal, of road and other transport in modern cities. Peter Calthorpe's article 'The Pedestrian Pocket' proposes dense inner-city development in which people walk to work rather than using transport, or use new light transit systems; in 'The Reconstruction of Social Meaning and the Space of Flows', Manuel Castells investigates information technology in relation to urban zones of affluence and poverty, arguing that the spaces of information (flows) will replace those of property (places) as the organising principles of power. Both essays are included in *The City Reader* (LeGates and Stout, 1996).

A well-illustrated descriptive text on the posters commissioned by London Underground is Oliver Green's *Underground Art* (Green, 1990). *Changing Stations* (LUL, 1993) sets out the policies and illustrates all the outcomes of the London Underground refurbishment programme of the 1980s. 'Art of the Line' (MBTA, 1986) and 'Travelling Hopefully' (Khan and Worpole, 1992) are reports on, respectively, art in the Boston T and art in public transport generally in the UK. The latter reads more as advocacy than enquiry.

A chapter in *Senseless Acts of Beauty* (McKay, 1996) – Direct Action of the new Protest: eco-rads on the road – locates road protest in context of the rise of direct action and single-issue politics, with a section on the resistance to the destruction of landscape for the extension of the M3 at Twyford Down, and one on resistance to the M11 at Leyton and Wanstead in East London, in which direct action was linked with performance art.

7 ART IN HEALTH SERVICES

The development of the asylum and hospital as institutions, reflecting the operations of control and discipline in post-Enlightenment society, is the subject of two works by Michel Foucault: *Madness and Civilisation* (Foucault, 1967) and *The Birth of the Clinic* (Foucault,

1973). These are a foundation for the critical consideration of art in health buildings today, a more interesting project than a scientific evaluation of the impact of art on recovery. A consideration of Rembrandt's *Anatomy Lesson* is given, in an essay on subjection, by Francis Barker in *The Tremulous Private Body* (Barker, 1984). Susan Sontag argued that there are punitive uses of illness in western culture in *Illness as Metaphor*, to which a sequel, *Aids and its Metaphors* was added in 1988 (published together, Sontag, 1990).

Amongst evaluative material, Roger Ulrich's 'View Through a Window Influences Recovery' from *Science* (Ulrich, 1984) is the most often encountered in discussion. Several other papers from medical journals are cited in footnotes to the chapter.

Art in health buildings receives minimal attention in the reports and advocacy of arts organisations, though the Arts Council's 'National Arts and Media Strategy' included a text on art and institutions by the writer.

Developments in the design of health facilities in the USA, generally more lavish than those of the UK, are covered by Jane Malkin, for example in *Medical and Dental Space Planning for the 1990s* (Malkin, 1990) and other publications. The NHS Executive issued two illustrated design guides to encapsulate its new design philosophy: *Environments for Quality Care* (HMSO, 1993) and *Environments for Quality Care: health buildings in the community* (HMSO, 1994a); the former was distributed to hospital estates departments, and the latter issued to every general practice in England.

8 ART AS A SOCIAL PROCESS

The case for an art of participation and ecological healing was put by Suzi Gablik in 1991 in *The Reenchantment of Art*. The argument takes on an added dimension if read in conjunction with Charlene Spretnak's *States of Grace*. For a more recent set of essays on the reconstruction of value in postmodernism, see *Principled Positions* (Squires, 1993), with contributions from Steven Connor, David Harvey, Paul Hirst, Chantal Mouffe, Christopher Norris, Kate Soper, Jeffrey Weeks and Iris Young.

Mapping the Terrain (Lacy, 1995) and *But is it Art?* (Felshin, 1995) have been previously mentioned, and *Culture in Action* (Jacob *et al.*, 1995) gives detailed accounts of all the projects included in Sculpture Chicago's 1993 season of alternatives to its previously conventional siting of sculpture in the city.

Two books give accounts of art and AIDS: *Don't Leave Me This Way* (Gott, 1994) was produced by the National Gallery of Australia to complement an exhibition of the same title, and gives descriptive and critical texts on a wide range of AIDS related art. *Living Proof* (Lowe and McMillan, 1992) was produced out of the residency (by the authors: a photographer and writer) in venues linked to AIDS in the north-east of England, and includes their work, that of people with whom they worked, and interviews.

Some of the positions of multiculturalism are rehearsed in Charles Taylor's *Multiculturalism* (Taylor, 1994), which includes an essay on 'Struggles for Recognition in the Democratic Constitutional State' by Habermas. A detailed account of the impact of white culture on native Australians in given in *Edge of Empire* by Jane M. Jacobs. For a comprehensive range of texts on post-colonialism, see *The Post-Colonial Studies Reader* (Ashcroft *et al.*, 1995), and for a black voice, see the writing of bell hooks.

A range of ecological art projects, including Mel Chin's *Revival Field*, is documented and illustrated in *Fragile Ecologies* (Matilsky, 1992). For a background of green politics, see *Green Political Thought* (Dobson, 1990) and *Modern Environmentalism* (Pepper, 1996). The essays in *Senseless Acts of Beauty* (McKay, 1996), for example on road protests, are again relevant, as is *Rainbow Nation Without Borders* (Buenfil, 1991), which charts the transition from the Woodstock generation to the communities and counter-cultures of the 1980s. For an introduction to 'deep ecology' see *Ecology, Community and Lifestyle* (Naess, 1990).

9 CONVIVIAL CITIES

For the research of W. H. Whyte, see *The Social Life of Small Urban Spaces* (Whyte, 1980) and *City* (Whyte, 1988). On the issues (and statistics) of urban development in the next century, including that of 'third world' cities, see: Akhtar Badshah's *Our Urban Future* (Badshah, 1996), *Sustainable Cities in Europe* by Peter Nijkamp and Adriaan Perrels (Nijkamp and Perrels, 1994), and Thomas Angotti's *Metropolis 2000* (Angotti, 1993). For an alternative, green perspective see Murray Bookchin's *Urbanization Without Cities* (Bookchin, 1992).

Some writing which reflects the optimism of the alternative cultures of the late 1960s, such as Victor Papanek's *Design for the Real World*, reprinted in 1991 (Papanek, 1991), and Ivan Illich's *Deschooling Society* and *Tools for Conviviality* (Illich, 1973a, 1973b), is worth re-reading in context of contemporary environmental discussion. So is *Where the Wasteland Ends* by Theodore Roszak (Roszak, 1973) which advocates a healing of the dualisms of reason–passion and mind–body, in an earlier and other form of the argument of Gablik's *The Reenchantment of Art*. The positions of winners of the Right Livelihood Awards (for creating modes of sustainable living) are given in two volumes: *Replenishing the Earth* (Woodhouse, 1990), and *Pioneers of Change* (Seabrook, 1993).

For a sequence of critical essays on society and consensus, see *Deconstruction and Pragmatism* (Mouffe, 1996); this may suggest a re-visiting of the discussion in Chapter 2 of representation and its support for active intervention rather than passive observation of difference. The collection includes Laclau's essay 'Deconstruction, Pragmatism, Hegemony'.

BIBLIOGRAPHY

•

Abramson, Cynthia (1995) 'Art and the Transit Experience', *Places*, vol. 9, no. 2, Summer 1995: 74–86.

Alexander, Christopher (1965) 'A City is not a Tree', *Architectural Forum*, vol. 122, no. 1, April 1965, in (eds) LeGates and Stout, 1996: 118–31.

Ambrose, Peter (1994) *Urban Process and Power*, London, Routledge.

Anderson, N. (1923) *The Hobo*, Chicago, University of Chicago Press.

Angotti, Thomas (1993) *Metropolis 2000*, London, Routledge.

Anon (1995) 'Art at Bay', *Art & Architecture*, no. 44, Spring: 10.

Apgar, Garry (1991) 'Public Art and the Remaking of Barcelona', *Art in America*, February 1991: 108–21.

Artists Agency (n.d. [1996]) 'Arts Health and Healing for Inner West End of Newcastle', Report to West End Health Resource Centre Trust, Newcastle.

Artists Agency (1996) *St Peter's Riverside Sculpture Project*, Sunderland.

Arts Council (1989) *An Urban Renaissance*, London.

Arts Council (1990) *Percent for Art – report of a steering group*, London.

Arts Council (1991) *Percent for Art : a review*, London.

The Art Studio (n.d.) publicity brochure, North Tyneside.

The Art Studio (1996) *Business Plan*, Sunderland.

Ashcroft, Bill, Griffiths, Gareth and Tiffin, Helen (eds) (1995) *The Post-Colonial Studies Reader*, London, Routledge.

Badshah, Akhtar (1996) *Our Urban Future*, London, Zed Books.

Baratloo, Mojdeh and Balch, Clifton (1989) *Angst: cartography*, New York, Lumen.

Barber, Stephen (1995) *Fragments of the European City*, London, Reaktion.

Barker, Francis (1984) *The Tremulous Private Body*, London, Methuen.

Barker, Francis (1993) *The Culture of Violence*, Manchester, Manchester University Press.

Baron, J. H., Hadden, M. and Palley, M. (1984) *St Charles Hospital Works of Art*, London, St Charles Hospital.

Baron, J. H. (1995) 'Art in Hospitals', Fitzpatrick Lecture 1994, *Journal of Royal College of Physicians of London*, vol. 29, no. 2, March: 131–44.

Barthes, Roland (1982) *Empire of Signs*, trans. Richard Howard, New York, Hill and Wang.

Batchelor, David (1995) 'Unpopular Culture', *Frieze*, no. 20, January.

Battersby, Christine (1995) 'Just Jamming: Irigaray, painting and psychoanalysis', in (ed.) K. Deepwell, *New Feminist Art Criticism*, Manchester, Manchester University Press.

Battersby, Jean (1994) 'Art and Airports', *Craft Arts International*, no. 30, 49–64.

Baudrillard, Jean (1994) *The Illusion of the End*, Cambridge, Polity.

Baxandall, Michael (1985) *Patterns of Intention*, London, Yale.

Beardsley, John (1989) *Earthworks and Beyond*, New York, Abbeville.

Becker, Carol (1994) *The Subversive Imagination*, New York, Routledge.

Benjamin, Andrew (1993) *The Plural Event*, London, Routledge.

Berman, Marshall (1983) *All That is Solid Melts into Air*, London, Verso.

Bertman, Sandra (1991) *Facing Death*, Bristol, Mass., Taylor and Francis.

Biggs, Lewis (1984) 'Open Air sculpture in Britain', in (eds) P. Davies and T. Knipe, *A Sense of Place*, Sunderland, Coelfrith: 13–39.

Billings, Jim (1990) 'Edgar Heap of Birds: A War Memorial', *Public Art Review*, vol. 2, no. 2, Fall/Winter 1990: 7.

Bird, Jon (1988) 'The Spectacle of Memory', in *Michael Sandle* (catalogue), London, Whitechapel Art Gallery: 29–39.

Bird, Jon, Curtis, Barry, Putnam, Tim, Roberston, George and Tickner, Lisa (eds) (1993) *Mapping the Futures*, London, Routledge.

Birmingham City Council (n.d.) *Art in the City*, sponsorship brochure for public art.

Bloch, Ernst [1959] (1986) *The Principle of Hope* (3 vols), trans. N. Plaice, S. Plaice and P. Knight, Oxford, Blackwell.

Blonsky, Marshall (1985) *On Signs*, Oxford, Blackwell.

Bloxham, Tom (1995) 'Regenerating the Urban Core', *Urban Design*, no. 53, January: 28–30.

Bookchin, Murray (1992) *Urbanization without Cities*, Montreal, Black Rose.

Borden, Ian, Kerr, Joe, Pivaro, Alicia and Rendell, Jane (eds) (1996) *Strangely Familiar*, London, Routledge.

Boyer, M. Christine (1986) *Dreaming the Rational City*, Cambridge, MA, MIT.

Breslow, Devra (1982) 'The Role of the Arts in US Cancer Care Centers', in Rockefeller Foundation, 1982: 30–43.

Brigham, Joan (1993) 'Reclamation Artists', *Leonardo*, vol. 26, no. 5: 379–85.

Brighton, Andrew (1993a) 'Flogging Dead Horses', *Things are not as they seem*, Canterbury, Kent Institute of Art and Design.

Brighton, Andrew (1993b) 'Is Architecture or Art the Enemy?', in (eds) de Ville and Foster (1993) 43–53.

British Rail (n.d.) *Glass Column*, London, British Rail Community Unit.

Broadgate (n.d.) *Vistors Guide*, London, Rosehaugh Stanhope.

Brown, Denise Scott (1990) *Urban Concepts*, London, Academy Editions.

Brown, Norman O., [1947] (1990) *Hermes the Thief*, Great Barrington, MA, Lindisfarne.

Brown, Norman O. (1966) *Love's Body*, New York, Random House.

Bryson, Norman, Holly, Michael, Ann and Moxey, Keith (eds) (1991) *Visual Theory*, Cambridge, Polity.

Buck-Morss, Susan (1995) *The Dialectics of Seeing: Walter Benjamin and the Arcades Project*, Cambridge, MA, MIT.

Buenfil, Alberto (1991) *Rainbow Nation Without Borders*, Santa Fe, Bear and Co.

Burgess, E. W., [1925] (1972) 'The Growth of a City', in (ed.) Stewart, 1972: 117–29.

Burnham, Linda F. (1989) 'Hands Across Skid Row: John Malpede's performance work-shop for the homeless of LA', in (ed.) Raven, 1989: 55–88.

Bynum, W. F., Porter R. and Shepherd M. (eds) (1985) *The Anatomy of Madness*, London, Tavistock.

Calvino, Italo, [1972] (1979) *Invisible Cities*, trans. W. Weaver, London, Pan.

Carde, Margaret (1990) 'Conversations with Earth', *Public Art Review*, vol. 2, no. 1, Spring/Summer 1990: 4–6.

Cardiff Bay Development Corporation (1990) *Strategy for Public Art in Cardiff Bay*, Cardiff, CBDC.

Carr, Stephen, Francis, Mark, Rivlin, Leanne and Stone, Andrew (1992) *Public Space*, Cambridge, Cambridge University Press.

Castells, Manuel (1983) *The City and the Grass Roots: a cross cultural theory of urban social mvements*, Berkeley, University of California Press.

Catchpole, Tim (1996) 'Agenda 21 in London', *Urban Design*, no. 57, January: 33–4.

de Certeau Michel, [1984] (1988) *The Practice of Everyday Life*, Berkeley, University of California Press.

Chambers, Iain (1993) 'Cities without Maps', in (eds) Bird *et al.*, 1993: 188–98.

Chrysler, Greig (1996) 'Making Places in Architectural History', in (ed.) King, 1996: 203–26.

Clarke, R. V. G. and Mayhew, P. (eds) (1980) *Designing Out Crime*, London, HMSO.

Coffield, Frank (1991) *Vandalism & Graffiti*, London, Gulbenkian.

Coles, Peter (1983) *Art in the National Health Service*, London, DHSS.

Coles, Peter (1985) *The Arts in a Health District*, London, DHSS.

Colomina, Beatriz (ed.) (1992) *Sexuality and Space*, New York, Princeton Architectural Press.

Conroy, Czrch and Litvinoff, Miles (1988) *The Greening of Aid*, London, Earthscan.

Coombes, Annie (1994) *Reinventing Africa*, New Haven, Yale.

Corbin, Alain (1986) *The Fragrant and the Foul: odor and the French imagination*, Cambridge, MA, Harvard University Press.

Cork, Richard (1987) 'Beyond the Tyranny of the Predictable', in *TSWA 3D*, Bristol, TSW and South West Arts Association: 6–13.

Cork, Richard (1988) 'From the Studio to the City', in Tess Jaray (catalogue), Serpentine Gallery, London: 29–34.

Cork, Richard (1989) 'For the Needs of Their Own Spirit: art in hospitals', in (ed.) Miles, 1989: 187–96.

Crawford, Margaret (1992) 'The World in a Shopping Mall', in (ed.) Sorkin, 1992: 3–30.

Cressey, P. G. (1932) *The Taxi-Hall Dancer*, Chicago, University of Chicago Press.

Curtis, Penelope (ed.) (1989) *Patronage and Practice – sculpture on Merseyside*, Liverpool, Tate Gallery and National Museums on Merseyside.

Dallas (1987) *Visual Dallas*, Dallas, Division of Cultural Affairs.

Darke, Jo (1991) *The Monument Guide*, London, Macdonald.

Davidoff, Paul (1965) 'Advocacy and Pluralism in Planning', *Journal of the American Institute of Planners*, vol. xxi, no. 4, November 1965; in (eds) LeGates and Stout, 1996: 422–32.

Davis, Kingsley (1965) 'The Urbanization of the Human Population', *Scientific American*, September 1965; reprinted in (eds) LeGates and Stout, 1996: 1–14.

Davis, Mike [1990] (1992) *City of Quartz*, London, Vintage.

Department of the Environment (1994) *Quality in Town and Country*, London, HMSO.

Department of Health & Social Security (DHSS) (1983) *Art in Hospitals*, conference report, London.

Department of Health & Social Security (DHSS) (1988) *Health Building Note 1*, London, HMSO.

Deutsche, Rosalyn (1991a), 'Uneven Development: public art in New York City', in (ed.) Ghirardo, 1991: 157–219.

Deutsche, Rosalyn (1991b), 'Alternative Space', in (ed.) Wallis, 1991: 45–66.

Digby, Anne (1985) 'Moral Treatment at the Retreat', in (eds) Bynum, Porter and Shepherd, 1985: 52–70.

Dillon, Wilton and Kotler, Neil (eds) (1994) *The Statue of Liberty Revisited*, Washington, DC, Smithsonian.

Dobbs, Margaret (1990) 'Alzheimer's Disease: the relationship between selected wall-covering patterns and resident behaviour in a special care unit', unpublished Ph.D. thesis, Texas Technical University.

Dobson, Andrew (1990) *Green Political Thought*, London, Routledge.

Dormer, Peter (1993a) *The Art of the Maker*, London, Thames and Hudson.

Dormer, Peter (1993b) 'Towards Better Design', *RA Magazine*, no. 38, Spring 1993: 39–40.

Dormer, Peter (1994) 'A Bridge too Far-sighted to be Built', *The Independent*, 23 March.

Doss, Erica (1995) *Spirit Poles and Flying Pigs*, Washington, DC, Smithsonian.

Duffin, Debbie (1992) 'Public Art as Analgesic', *Public Art Review*, Summer, 1992: 3, 26.

Duncan, James S. (1996) 'Me(trope)olis: or, Hayden White among the urbanists', in (ed.) King, 1996: 253–68.

Duncan, Nancy (ed.) (1996) *Body Space*, London, Routledge.

Dunn, Peter and Leeson Loraine (1993) 'The Art of Change in Docklands', in (eds) Bird *et al.*, 1993: 136–49.

Eliot, T. S., [1917] (1971) *The Wasteland – facsimile and transcript*, London, Faber.

Elsen, Albert (1985) *Rodin's Thinker*, New Haven, CT, Yale.

Engberg, Kristen (1991) 'Art Against AIDS', *New Art Examiner*, May 1991: 25.

Erikson, Erik (1959) 'The Problem of Ego Identity', *Psychology Issues*, vol. 1, no. 1: 102.

Fanon, Frantz, [1961] (1990) *The Wretched of the Earth*, London, Penguin.

Farris, James C. (1972) *Nuba Personal Art*, London, Duckworth.

Featherstone, Mike (1995) *Undoing Culture*, London, Sage.

Federal Transit Authority (FTA) (1996a) *Art in Transit*, Washington.

Federal Transit Authority (FTA) (1996b) *Art in Transit: making it happen*, Washington.

Felshin, Nina (ed.) (1995) *But is it Art?*, Seattle, Bay Press.

Feuer, Wendy (1993) 'Public Art from a Public Sector Perspective', in (ed.) Raven, 1993: 139–54.

Filochowski, Jan and Cadenhead, Neil (1995) 'Cultivating Change', *Hospital Development*, January: 22–5.

Fisher, Mark and Rogers, Richard (1992) *A New London*, London, Penguin.

Fishman, Robert (1987) *Bourgeois Utopias*, New York, Basic Books.

Flanagan, Regina and Jones, Seitu (1991) 'A dialogue on Lucy Lippard's Mixed Blessings', *Public Art Review*, vol. 3, no. 2, Fall/Winter: 4–5.

Fleming, Ronald and von Tscharner, Renata (1987) *Placemakers*, Boston, Harcourt Brace Jovanovich.

Forester, John (1987a) *Critical Theory and Public Life*, Cambridge, MA, MIT.

Forester, John (1987b) 'Planning in the Face of Conflict', *Journal of the American Planning Association*, vol. 53, no. 3, Summer, in (eds) LeGates and Stout, 1996, 433–46.

Foster, Hal (ed.) (1983) *The Anti-Aesthetic*, Seattle, Bay Press.

Foster, Hal (1985) *Recodings*, Seattle, Bay Press.

Foster, Hal (ed.) (1987) *Discussions in Contemporary Culture*, Seattle, Bay Press.

Foster, Hal (ed.) (1988) *Vision and Visuality*, Seattle, Bay Press.

Foster, Richard (1991) *Patterns of Thought*, London, Cape.

Foucault, Michel (1967) *Madness and Civilisation*, trans. R Howard, London, Tavistock.

Foucault, Michel (1973) *The Birth of the Clinic*, trans. A. M. Sheridan, London, Tavistock.

Freshman, Phil (1992) *Public Address*, Minneapolis, Walker Art Centre.

Frisby, David (1985) *Fragments of Modernity*, Cambridge, Polity.

Frisby, David (1992) *Simmel and Since*, London, Routledge.

Fuller, Peter (1980) *Beyond the Crisis in Art*, London, Readers and Writers Cooperative.

Fuller, Peter (1985) *Images of God*, London, Chatto.

Fuller, Peter, [1980] (1988) *Art and Psychoanalysis*, London, Hogarth Press.

Gablik, Suzi (1991) *The Reenchantment of Art*, London, Thames and Hudson.

Gablik, Suzi (1993) 'Art to Nurture the World', *Resurgence*, no. 156, Jan/Feb 1993: 28–9.

Gablik, Suzi (1995) 'Connective Aesthetics: art after individualism', in (ed.) Lacy, 1995: 74–87.

Gare, Aaren (1995) *Postmodernism and Environmental Crisis*, London, Routledge.

Gass, William (n.d.) 'Monument/Mentality', in (ed.) K. Foster, *Oppositions*, Fall: cited in Warner (1985) *Monuments and Maidens*, London, Picador: 12, 338.

Gayle, Margot and Cohen Michèle (1988) *Manhattan's Outdoor Sculpture*, New York, Prentice Hall.

Ghirardo, Diane (ed.) (1991) *Out of Site*, Seattle, Bay Press.

Ghirardo, Diane (1995) 'The Case for Letting Malibu Burn and Slide', in (ed.) Lang, 1995: 97–101.

Girardet, Herbert (1992) *The Gaia Atlas of Cities*, London, Gaia: 19.

Gold, Seymour M. (1977) 'Social Benefits of Trees in Urban Environments', *Journal of Environmental Studies*, vol. 10: 85–90.

Goody, Joan (1993) 'From Project to Community: the redesign of Columbia Point', *Places*, vol. 8, no. 4, Summer 1993: 20–33.

Gott, Ted (ed.) (1994) *Don't Leave Me This Way – art in the age of AIDS*, Canberra, National Gallery of Australia.

Granaat, David (1982) 'The Plastic Arts in Hospitals', in Rockefeller Foundation, 1982: 97–107.

Gratz, Roberta (1989) *The Living City*, New York, Simon and Schuster.

Greater London Council (1986) *Campaign for a Popular Culture*, London.

Green, Lynne (1993) 'Pleasing the earth', *Contemporary Art*, vol. 1, no. 4, Summer: 36–40.

Green, Oliver (1990) *Underground Art*, London, Studio Vista.

Greene, John (1987) *Brightening the Long Days*, London, Tiles and Architectural Ceramics Society.

Greene, Lesley (1987) 'The Longest Art Gallery in the World', in (ed.) Petherbridge, 1987: 56–7.

Griswald, Charles (1992) 'The Vietnam Veterans' Memorial', in (ed.) Mitchell, 1992: 79–112.

Grunenberg, Christopher (1994) 'The Politics of Presentation: the Museum of Modern Art, New York', in (ed.) Pointon, 1994: 192–211.

Guilbaut, Serge (1983) *How New York Stole the Idea of Modern Art*, Chicago, University of Chacago Press.

Hargreaves, David (1982) *The Challenge for the Comprehensive School*, London, Routledge.

Harris, Stacey Paleologos (1984) *Insights/On Sites*, Washington, DC, Partners for Livable Places.

Harvey, David (1989) *The Urban Experience*, Baltimore, Johns Hopkins University Press.

Harvey, David (1993) *From Space to Place and Back Again*, in (eds) Bird *et al.*, 1993: 3–29.

Hayden, Dolores (1995) *The Power of Place*, Cambridge, MA, MIT.

Health Care Arts (1994) *Crafts in Scottish Hospitals*, Dundee (not paginated, loose-leaf folio).

Heartney, Eleanor (1992) 'Art and its Audience: a public experiment', in (ed.) Shamash, 1992: 17–21.

Heartney, Eleanor (1995) 'Ecopolitics/Ecopoetry: Helen and Newton Harrison's environmental talking cure', in (ed.) Felshin, 1995: 141–63.

Heath, Jane (ed.) (1992) *The Furnished Landscape*, London, Bellow Publishing (for the Crafts Council).

Heerwagen, Judith H. (1990) 'The Psychological Aspects of Windows and Window Design', *Proceedings of the Environmental Design Research Association*, vol. 21: 269–80.

Heidegger, Martin, [1950] (1975) *Early Greek Philosophy*, trans. D. F. Krell and F. A. Capuzzi, San Francisco, Harper and Row.

Heidegger, Martin, [1971] (1975) *Poetry, Language, Thought*, trans. A Hofstadter, New York, Harper and Row.

Hewison, Robert (1987) *The Heritage Industry*, London, Methuen.

Hillman, James and Ventura, M. (1992) *We've had 100 Years of Psychotherapy and the World's Still Getting Worse*, San Francisco, Harper and Row.

Holman, Valerie (1991) 'Evaluating Art in Hospitals: a report based on research in Oxford Hospitals', unpublished report for Health Care Arts.

Holston, Bill (1992) *Drawing Support*, Belfast, Beyond the Pale.

Holub, Renate (1992) *Antonio Gramsci – beyond Marxism and Postmodernism*, London, Routledge.

Holub, Robert (1991) *Jurgen Habermas – critic in the public sphere*, London, Routledge.

hooks, bell (1995) *Art on my Mind*, New York, The New Press.

Horne, Donald (1986) *The Public Culture*, London, Pluto.

Hoy, David Couzens, and McCarthy, Thomas (1994) *Critical Theory*, Oxford, Blackwell.

Hughes, Graham (1988) 'On the Rails', *Arts Review Year Book* (reprinted by British Rail).

Hunt, Lynn (1989) *The New Cultural History*, Berkeley, University of California Press.

Illich, Ivan [1971] (1973a), *Deschooling Society*, London, Penguin.

Illich, Ivan (1973b), *Tools for Conviviality*, London, Calder and Boyars.

Illich, Ivan (1976) *Limits to Medicine*, London, Maryoor Boyars.

Illich, Ivan [1984] (1986) *H₂0 and the Waters of Forgetfulness,* London, Boyars.

Irigaray, Luce (1993) *Sexes and Genealogies*, New York, Columbia University Press.

Irigaray, Luce (1994) *Thinking the Difference*, trans. Karin Montin, Dublin, Athlone.

Jackson, J. B. (1972) *American Space*, New York, Norton.

Jackson, J. B. (1980) *The Necessity for Ruins*, Amherst, University of Massachusetts Press.

Jackson, Peter (1989) *Maps of Meaning*, London, Routledge.

Jacob, Mary Jane (1996) 'Press Release', for *Conversation in the Castle*.

Jacob, Mary Jane, Brenon, Michael and Olson, Eva (1995) *Culture in Action*, Seattle, Bay Press.

Jacobs, Allen and Appleyard, Donald (1987) 'Toward an Urban Design Manifesto', *Journal of the American Planning Association*, vol. 53, no. 1, Winter: in (eds) LeGates and Stout, 1996, 165–75.

Jacobs, Jane (1961) *The Death and Life of Great American Cities*, New York, Random House.

Jacobs, Jane M. (1996) *Edge of Empire*, London, Routledge.

Jencks, Charles (1984) 'A Modest Proposal on the Collaboration Between Artist and Architect', in (ed.) Townsend, 1984: 14–19.

Jones, Susan (ed.) (1992) *Art in Public*, Sunderland, AN Publications.

Jones, Tony Lloyd (1996) 'Curitiba: sustainability by design', *Urban Design*, no. 57, January: 26–32.

Kaiser, Leland (n.d.) 'The Hospital as a Healing Community', unpublished conference paper.

Kaprow, Allan (1993) *The Blurring of Art and Life* (ed. J Kelley), Berkeley, University of California Press.

Karanian, William (1994) 'Tree of Learning', *Hospital Development*, November: 25–9.

Karp, Ivan, Kreamer, Christine and Lavine, Steven (eds) (1992), *Museums and Communities – the politics of public culture*, Washington, Smithsonian.

Kastner, Jeff (1995) 'Art as a Verb', *Artists Newsletter,* April: 24–5.

Keep, P., James J. and Inman, M. (1980) 'Windows in the Intensive Therapy Unit', *Anaesthesia*, vol. 35, 257–62.

Kerenyi, Karl [1944] (1986) *Hermes, Guide of Souls*, trans. Murray Stein, New York, Spring.

al-Khalil, Samir (1991) *The Monument*, Berkeley, University of California Press.

Khan, Nasim and Worpole, Ken (1992) *Travelling Hopefully*, London, Report for the Gulbenkian Foundation (UK).

Kiepper, Alan (1994) 'Art: another way to move people', *Public Transport International*, no. 2: 15–19.

King, Anthony (ed.) (1996) *Re-Presenting the City*, London, Macmillan.

King Edward's Hospital Fund for London (1977), *Psychiatric Hospitals Viewed by Their Patients*, London.

Kluge, Alexander (1981–2) 'On film and the public sphere', *New German Critique*, 24–5, Fall/Winter: 212.

Kruger, Barbara and Mariani, Phil (eds) (1989) *Remaking History*, Seattle, Bay Press.

Kuspit, Donald (1989) 'Notes on American Activist Art Today', in (ed.) Raven (1989) [1993]: 255–68.

Lacy, Suzanne (1992) 'Does Art Heal?', unpublished lecture transcript, Conference of the Society of Healthcare Arts Administrators, The Dalles, USA, October 1992.

Lacy, Suzanne (1993) 'Fractured Space', in (ed.) Raven, 1993: 287–302.

Lacy, Suzanne (ed.) (1995) *Mapping the Terrain*, Seattle, Bay Press.

Landry, Charles and Bianchini, Franco (1995) *The Creative City*, London, Demos.

Lang, Peter (1995a) *Fear Not*, interview with Mark Wigley, in (ed.) Lang, 1995b: 4–7.

Lang, Peter (ed.) (1995b) *Mortal City*, New York, Princeton Architectural Press.

Lavenda, Robert (1992) 'Festivals and the Creation of Public Culture: whose voice(s)?', in (eds) Karp, Kreamer and Lavine, 1992: 76–104.

Leavis, F. R., [1932] (1976) *New Bearings in English Poetry*, London, Penguin.

Le Corbusier [1929] (1987) *The City of Tomorrow*, trans. F. Etchells, New York, Dover.

Lefebvre, Henri [1974] (1991) *The Production of Space*, trans. D. Nicholson-Smith, Oxford, Blackwell.

Lefebvre, Henri [1947] (1992) *Critique of Everyday Life*, trans. J. Moore, London, Verso.

Lefebvre, Henri (1996) *Writings on Cities*, trans. E. Kofman and E. Lebas, Oxford, Blackwell.

LeGates, Richard and Stout, Frederic (eds) (1996) *The City Reader*, London, Routledge.

Leger, Fernand, [1924] (1970) 'The Machine Aesthetic, The Manufactured Object, The Artisan and the Artist', *Bulletin de l'Effort Moderne* 1 and 2, reproduced in exhibition catalogue *Leger and Purist Paris*, London, Tate Gallery: 87–92.

Lingwood, James (ed.) (1996) *House*, London, Phaidon.

Lippard, Lucy (1989) 'Moving Targets/Moving Out' in (ed.) Raven, 1989: 209–28.

Lippard, Lucy (1990) *Mixed Blessings: New Art in a Multicultural America*, New York, Pantheon.

Loftman, Patrick and Nevin, Brendan (1992) 'Urban Regeneration and Social Equity: a case study of Birmingham', research paper 8, Birmingham, University of Central England (Faculty of the Built Environment).

London Docklands Development Corporation (LDDC) (1991) *London Docklands Arts*, London, LDDC.

London Underground Ltd (LUL) (1993) *Changing Stations*, London, LUL.

Lowe, Nicholas and McMillan, Michael (1992) *Living Proof*, Sunderland, Artists Agency.

Lurie, David and Wodiczko, Krzysztof (1988) 'The Homeless Vehicle Project', *October*, no. 47: 53–67.

Lynch, Kevin (1960) *The Image of the City*, Cambridge, MA, MIT.

Lynch, Kevin, [1981] (1984) *Good City Form*, Cambridge, MA, MIT.

McAvera, Brian (n.d.) [1990?] *Art, Politics and Ireland*, Dublin, Open Air.

Mackay, David (1991) 'Urban Design and the Cultural Interface', *Urban Design Quarterly*, no. 37, January: 5–9.

McEwen, Indra Kagis (1993) *Socrates' Ancestor*, Cambridge, MA, MIT.

McEwen, John (1984) 'An Alternative Avant-Garde: the role of community arts', in (ed.) Townsend, 1984: 72–7.

MacLachlan, Gale and Reid, Ian (1994) *Framing and Interpretation*, Melbourne, Melbourne University Press.

McLaughlin, Corinne and Davidson, Gordon (1985) *Builders of the Dawn*, Summertown, TN, Book Publishing Company.

Malkin, Jane (1990) *Medical and Dental Space Planning for the 1990s*, New York, Van Nostrand Reinhold.

Marcuse, Herbert, [1937] (1972) 'The Affirmative Character of Culture', in H. Marcuse, 1972, *Negations*, London, Penguin: 88–133.

Marcuse, Herbert (1978) *The Aesthetic Dimension*, Boston, MA, Beacon Press.

Marcuse, Herbert [1956] (1987) *Eros and Civilisation*, London, Routledge and Kegan Paul.

Martz, Louis L. (1962) *The Poetry of Meditation*, New Haven, CT, Yale.

Massey, Doreen (1993) 'Power-geometry and a progressive sense of place' in (eds) Bird *et al.*, 1993: 59–69.

Massey, Doreen (1994) *Space, Place and Gender*, Cambridge, Polity.

Massey, Doreen (1995) 'Together Architects and Planners must Integrate the Spheres of Work, Leisure and Home', *The Independent*, 22 March.

Matilsky, Barbara (1992) *Fragile Ecologies*, New York, Rizzoli.

Massachusetts Bay Transportation Authority (MBTA) (1986) 'Arts on the line: eight year report, Boston, MA, MBTA.

Meller, Helen (1979) *The Ideal City*, London, Unwin.

Meyer, Richard (1995) 'This is to Enrage You', in (ed.) Felshin, 1995: 51–82.

Miles, Malcolm (ed.) (1989) *Art for Public Places*, Winchester, Winchester School of Art Press.

Miles, Malcolm (1991a) 'Is There Still Life Amongst the Ruins?', *Journal of Art & Design Education*, vol. 10, no. 2: 159–65.

Miles, Malcolm (1991b) 'The Tree of Learning', *Hospital Development*, November: 21–3.

Miles, Malcolm (1992) 'Altered Image', *Hospital Development*, January: 20–3.

Miles, Malcolm (1994) 'Art in Hospitals: does it work?', *Journal of the Royal Society of Medicine*, vol. 87, March: 161–3.

Miles, Malcolm (ed.) (1995a) *Art and the City*, Portsmouth, University of Portsmouth Press.

Miles, Malcolm (1995b) 'Art and Urban Regeneration', *Urban History*, vol. 22, August: 71–85.

Miles, Malcolm (1995c) 'Room with a View – whiteness', *Point*, no. 1, Winter: 53–8.

Miller, Cheryl (1990) 'Out of the Trashcan', *Public Art Review*, vol. 2, no. 1, Spring/Summer: 7.

Mitchell, Juliet (ed.) (1986) *Selected Melanie Klein*, London, Penguin.

Mitchell, W. J. T. (1992) *Art and the Public Sphere*, Chicago, University of Chicago Press.

Mitchell, William (1996) *City of Bits*, Cambridge, MA, MIT.

Mollenkopf, John (1992) *A Phoenix in the Ashes*, Princeton, NJ, Princeton University Press.

Molyneux, John (1985) 'A Modest Proposal for Portsmouth', in (ed.) Miles, 1995a: 16–18.

Montgomerey, John (1992) 'Dressed to Kill Off Urban Culture', *Planning*, 9 October: 6–7.

More, David (1984) 'The Buffalo Underground', *Art Monthly*, June: 7–9.

Morley, Henry (1887) *Ideal Commonwealths*, London, George Routledge and Sons.

Mouffe, Chantal (ed.) (1996) *Deconstruction and Pragmatism*, London, Routledge.

Mourey, Gabriel (1904) in *Les Arts de la Vie*, June; cited in Elsen, 1985.

Mozingo, Louise (1989) 'Women and Downtown Open Spaces', *Places*, vol. 6, no. 1, Fall: 38–47.

MTA (1987) *Annual Report*, New York.

MTA (1994) *Art en Route*, New York.

MTA (n.d.) *MTA ARTS for Transit Permanent Art Program*, New York.

Mumford, Lewis, [1961] (1966) *The City in History*, London, Penguin.

Myerscough, John (1988) *The Economic Importance of the Arts in Britain*, London, Policy Studies Institute.

Naess, Arne (1990) *Ecology, Community and Lifestyle*, trans. D. Rothenberg, Cambridge, Cambridge University Press.

National Endowment for the Arts (NEA) (1981) *Art in Public Places*, Washington, DC, Partners for Livable Places.

National Health Service Executive (NHSE) (1991) *Demonstrably Different*, London, HMSO.

National Health Service Executive (NHSE) (1992) *First Impressions, Lasting Quality*, London, HMSO.

National Health Service Executive (NHSE) (1993) *Environments for Quality Care*, London, HMSO.

National Health Service Executive (NHSE) (1994a) *Environments for Quality Care: health buildings in the community*, London, HMSO.

National Health Service Executive (NHSE) (1994b) *Changing Perspectives*, London, HMSO.

Ngo, Viet (n.d.) Publicity Material, St Paul, Minnesota, Lemna Corporation.

Nijkamp, Peter and Perrels, Adriaan (1994) *Sustainable Cities in Europe*, London, Earthscan.

O'Hear, Anthony (1994) 'What is Aesthetic Value', in *Artists in the 1990s*, London, Wimbledon School of Art and the Tate Gallery: 59-65.

Overy, Paul (1986) 'Paolozzi's Underground: art at work', *Art Monthly*, March: 12–13.

Owens, Craig (1983) 'The Discourse of Others: feminists and postmodernism', in (ed.) Foster, 1983: 71.

Palmer, Janice, and Nash, Florence (1993) 'Taking Shape', *North Carolina Medical Journal*, vol. 54 no. 2, February: 101–4.

Papanek, Victor, [1984] (1991) *Design for the Real World*, London, Thames and Hudson

Park, R. E., [1915] (1967) 'The City: suggestions for the investigation of human behaviour in the city environment', in (eds) R. E. Park *et al.* (1967) *The City*, Chicago, University of Chicago Press.

Patterson, Ian (1989) 'Evaluation Report on the Artist in Residence and Musician at Aycliffe Hospital', unpublished report, Newcastle, Northern Regional Health Authority.

Pearman, Hugh (1991) 'Putting the Art Back into Architecture', *Sunday Times*, 3 February.

Peloquin, Suzanne (1991) 'Art in Practice: when art becomes caring', unpublished Ph.D. thesis, University of Texas at Galveston.

Pepper, David (1996) *Modern Environmentalism*, London, Routledge.

Petherbridge, Deanna (ed.) (1987) *Art for Architecture*, London, HMSO.

Petherbridge, Deanna (1994) 'Passionate and Dispassionate Patronage', in (ed.) Townsend, 1994: 20–9.

Phillips, Patricia (1988) 'Out of Order: the public art machine', *Artforum*, December: 93–6.

Phillips, Patricia (1989) 'Introduction', to Baratloo and Balch, 1989: 5–12.

Phillips, Patricia (1994a) 'The Private is Public', *Public Art Review*, Spring/Summer: 16–17.

Phillips, Patricia (1994b) 'Landscape as Subject and Medium', *Public Art Issues*, no. 3: 11–13.

Phillips, Patricia (1995a) 'Maintenance Activity: creating a climate for change', in (ed.) Felshin, 1995: 165–94.

Phillips, Patricia (1995b) 'Peggy Diggs: private acts and public art', in (ed.) Felshin, 1995: 283–308.

Pinkus, Richard, 'The Invisible Town Square', in (ed.) Felshin, 1995: 31–49.

Plunz, Richard (1995) 'Beyond Dystopia, Beyond Theory Formation', in (ed.) Lang, 1995: 28–36.

Pointon, Marcia (ed.) (1994) *Art Apart*, Manchester, Manchester University Press.

Pollock, Griselda (1985) *Vision and Difference*, London, Routledge.

Proudfoot, Peter (1996) 'Government Control in Urban Waterfront Renewal', *Urban Design*, vol. I, no. 1, February: 105–14.

Public Art Commissions Agency (1990) *Context & Collaboration*, conference papers, Birmingham, PACA.

Public Art Fund (PAF) (1994) *Urban Paradise: gardens in the city*, New York, PAF.

Public Art Fund (PAF) (1996) *In process*, quarterly newsletter, vol. 4, no. 2, Winter, PAF.

Pym, William (1995) *St Peter's Riverside Sculpture Project*, evaluation report (unpublished).

Raven, Arlene [1989] (1993) *Art in the Public Interest*, New York, da Capo.

Raymond, Marcel [1933] (1970) *From Baudelaire to Surrealism*, London, Methuen.

Reiter, Wellington (1995) 'Bridges and Bridging: infrastructure and the arts', *Places*, vol. 9, no. 2, Summer: 60–7.

Relf, Diane (ed.) (1990) *The Role of Horticulture in Human Well-Being* (papers from a symposium at Arlington, Virginia), Portland, Timber Press.

Richards, Brian (1990) *Transport in Cities*, London, Architecture, Design and Technology Press.

Rilke, Rainer Maria [1910] (1988) *The Notebooks of Malte Laurids Brigge*, New York, Vintage.

Roberts, Marion Salter, Miffa and Marsh, Chris (1993) *Public Art in Private Places*, London, University of Westminster Press.

Rockefeller Foundation (1978) *The Healing Role of the Arts*, New York, Rockefeller Foundation.

Rockefeller Foundation (1982) *The Healing Role of the Arts: a European Perspective*, New York, Rockefeller Foundation.

Rodgers, Peter (1989) *The Work of Art*, London, Gulbenkian.

Rogers, Richard (1995) 'Private Greed and Public Responsibility', *Art & Architecture*, no. 43, Autumn 1995, 2–11.

Rose, Gillian (1993) 'Some Notes Towards Thinking About the Spaces of the Future', in (eds) Bird *et al.*, 1993: 70–86.

Rosler, Martha (1991) 'Fragments of a Metropolitan Viewpoint', in (ed.) Wallis, 1991: 15–44.

Rosler, Martha (1994) 'Place, Position, Power, Politics', in (ed.) Becker, 1994: 55–76.

Ross, Nick (1994) 'Proceedings of the RSA', *RSA Journal*, November: 44–9.

Rosser, Phyllis (1994) 'Mel Chin's Invisible Architecture', *Public Art Review*, vol. 6, no. 1, Fall/Winter: 30.

Roszak, Theodore (1973) *Where the Wasteland Ends*, London, Faber.

Roth, Moira (1989) 'Suzanne Lacy: social reformer and witch', in (ed.) Raven, 1989: 155–73

Rowe, Colin and Koetter, Fred (1978) *Collage City*, Cambridge, MA, MIT.

Rowland, Jon (1995) 'The Urban Design Process', *Urban Design*, no. 56, October: 22–7.

Rubik, Beverly (ed.) (1992) *The Interrelationship Between Mind and Matter*, conference papers, Philadelphia, Temple University.

Rupp, James (1992) *Art in Seattle's Public Places*, Seattle, University of Washington Press.

Russell, Bertrand, [1946] (1961) *History of Western Philosophy*, London, Unwin Hyman.

Rykwert, Joseph, [first edn 1976] (1988) *The Idea of a Town*, Cambridge, MA, MIT.

Sacks, Shelley (1995) 'Joseph Beuys' Pedagogy and the Work of James Hillman: the healing of art and the art of healing', *Issues*, Winter: 52–60.

Said, Edward (1994) *Culture and Imperialism*, London, Vintage.

Sassen, Saskia (1991) *The Global City*, Princeton, NJ, Princeton University Press.

Sassen, Saskia (1996) 'Analytic Borderlands' in (ed.) King, 1996: 183–202.

Savage, Mike and Warde, Alan (1993) *Urban Sociology, Capitalism and Modernity*, London, Macmillan.

Savitch, H. V. (1988) *Post-Industrial Cities*, Princeton, NJ, Princeton University Press.

Schroder-Sheker, Therese (1993) 'Music for the Dying', *Advance*, vol. 9, no. 1, Winter: 36–48.

Seabrook, Jeremy (ed.) (1993) *Pioneers of Change*, London, Zed Books.

Selwood, Sara (1991) 'Public Art, Private Amenities', *Art Monthly*, February: 1–3.

Selwood, Sara (1992) 'Art in Public', in (ed.) Jones, 1992: 11–27.

Selwood, Sara (1995) *The Benefits of Public Art*, London, PSI.

Sennett, Richard (1996) *The Uses of Disorder*, London, Faber.

Sennett, Richard [1990] (1992) *The Conscience of the Eye*, New York, Norton.

Sennett, Richard [1974] (1992) *The Fall of Public Man*, New York, Norton.

Sennett, Richard (1994) *Flesh and Stone*, London, Faber.

Shamash, Diane (ed.) (1991) *A Field Guide to Seattle's Public Art*, Seattle, Seattle Arts Commission.

Shamash, Diane (ed.) (1992) *In Public: Seattle 1991*, Seatle, Seattle Arts Commission.

Shields, Rob (1996) 'Alternative Traditions of Urban Theory', in (ed.) King, 1996: 227–52.

Short, John Rennie (1991) *Imagined Country*, London, Routledge.

Sibley, David (1995) *Geographies of Exclusion*, London, Routledge.

Simmel, Georg [1902] (1964) 'The Metropolis and Mental Life', in (ed.) Wolff, K. M., 1964: 409–24.

Smith, Michael Porter (1980) *The City and Social Theory*, Oxford, Blackwell.

Smith, Neil (1993) 'Homeless/global: scaling places', in (eds) Bird *et al.*, 1993: 87–119.

Soja, Edward (1989) *Postmodern Geographies*, London, Verso.

Soja, Edward (1992) 'Inside Exopolis: scenes from Orange County', in (ed.) Sorkin, 1992: 94–122.

Sonfist, Alan (ed.) (1983) *Art in the Land*, New York, Dutton.

Sontag, Susan (1990) *Illness as Metaphor and Aids and its Metaphors*, New York, Doubleday.

Sorkin, Michael (ed.) (1992) *Variations on a Theme Park*, New York, Noonday.

South Tees Health (n.d.) [1989?], *Putting People First*, internal paper for managers in South Tees Hospitals, Middlesbrough.

Speer, Albert (1970) *Inside the Third Reich*, New York, Weidenfeld and Nicholson; cited in Samir al-Khali, 1991: 39.

Spencer, Michael Jon (1978) 'A Case for the Arts', in Rockefeller Foundation, 1978: 1–9.

Spitzer, Theodore and Baum, Hilary (1995) *Public Markets and Community Revitalization*, New York, Projects for Public Spaces and Urban Land Institute.

Spretnak, Charlene, [1991] (1993) *States of Grace*, New York, Harper Collins.

Squires, Judith (ed.) (1993) *Principled Positions*, London, Lawrence and Wishart.

Stewart, Murray (ed.) (1972) *City*, London, Penguin.

Storr, Robert (1991) *Dislocations*, catalogue essay, New York, Museum of Modern Art: 18–33.

Taylor, Brandon (1994) 'From the Penitentiary to the Temple of Art: early metaphors of improvement at the Millbank Tate', in (ed.) Pointon, 1994: 9–32.

Taylor, Charles (1994) *Multiculturalism*, Princeton, NJ, Princeton University Press.

Taylor, Paul (1992) 'Art on the Barricades', *Outrage*, no. 113, October: 29.

Thamesdown Borough Council (1988) *Art in the Urban Environment*, Swindon.

Tickner, Lisa (1987) *Nancy Spero*, exhibition catalogue, London, ICA.

Tilyard, E. M. W., [1943] (1972) *The Elizabethan World Picture*, London, Penguin.

Tolstoy, Vladimir, Bibikova, Irina and Cooke, Catherine (1990) *Street Art of the Revolution*, London, Thames and Hudson.

Townsend, Peter (ed.) (1984) *Art Within Reach*, London, Thames and Hudson.

Trodd, Colin (1994) 'Culture, Class, City: the National Gallery, London and the spaces of education, 1822–57', in (ed.) Pointon, 1994: 33–49.

Ulrich, Roger (1984) 'View Through a Window May Influence Recovery', *Science*, no. 224: 224–5.

Ulrich, Roger (1991) 'Effects of Interior Design on Wellness', *Journal of Health Care Interior Design*, 3: 97–109.

Urban Design Group (UDG) (1987) Agenda, in UDG, 1994: 34–5.

Urban Design Group (UDG) (1994) *Urban Design Source Book*, Oxford, UDG.

Urban, Thresholds (1989) *Art Meets Architecture in the City*, Perth, Urban Thresholds.

Urry, John (1990) *The Tourist Gaze*, London, Sage.

Urry, John (1995) *Consuming Places*, London, Routledge.

Vance, Carole (1994) 'The War on Culture', in (ed.) Gott, 1994: 91–111.

de Ville, Nicholas and Foster, Stephen (eds) (1993) *Space Invaders*, Southampton, John Hansard Gallery.

Wallis, Brian (ed.) (1991) *If You Lived Here*, Seattle, Bay Press.

Walton, J. K. (1985) 'Casting Out and Bringing Back in Victorian England: pauper lunatics', in (eds) Bynum, Porter and Shepherd, 1985: 132–46.

Warner, Marina (1985) *Monuments and Maidens*, London, Picador.

Watney, Simon (1994) 'Art From the Pit: some reflections on monuments, memory and AIDS', in (ed.) Gott, 1994: 52–62.

Wates, Nick (1996) 'A Community Process', *Urban Design*, conference proceeding, April: 15–17.

Weinstein, Jeff (1989) 'Names Carried into the Future: an AIDS quilt unfolds', in (ed.) Raven, 1989: 43–54.

Wenzel, Marian (1972) *House Decoration in Nubia*, London, Duckworth.

Western Australia (1988) *Art for Liveable Places*, Perth, W. A. Department for the Arts.

Western Australia (1990) *A Strategy for Public Art*, Perth, W. A. Department for the Arts.

White, Jerry (1981) 'Campbell Road: the worst street in north London', *History Today*, vol. 31, June, 5–9.

White, John [1957] (1972) *The Birth and Rebirth of Pictorial Space*, London, Faber.

Whyte, William H. (1980) *The Social Life of Small Urban Spaces*, Washington, DC, Conservation Foundation.

Whyte, William H. (1988) *City: rediscovering the centre*, New York, Doubleday.

Widgery, David (1991) *Some Lives! A GP's East End*, London, Sinclair-Stevenson.

Wigley, Mark (1992) 'Untitled – the housing of gender', in (ed.) Colomina, 1992: 327–89.

Wigley, Mark (1995) 'Interview with Peter Lang', see Lang, 1995: 70–81.

Willett, John (1967) *Art in a City*, London, Methuen.

Willett, John (1984) 'Back to the Dream City: the current interest in public art', in (ed.) Peter Townsend, 1984: 6–13.

Williams, Gwyn (1976) *Goya and the Impossible Revolution*, London, Allen Lane.

Wilson, Elizabeth (1991) *The Sphinx and the City*, Berkeley, University of California Press.

Wilson, Larkin (1972) 'Intensive Care Delirium', in *Arch. Intern. Med.*, vol. 130, August: 225–6.

Winkel, Gary and Holohan, Charles (1985) 'The Environmental Psychology of the Hospital: is the cure worse than the illness?', *Prevention in Human Services*, part 4: 11–33.

Wirth, Louis (1938) 'Urbanism as a Way of Life', *American Journal of Sociology*, XLIV, 1: 1–24.

Wodiczko, Krzysztof (1991) *The Homeless Vehicle Project*, Kyoto, Kyoto Shoin International.

Wolff, Janet (1981) *The Social Production of Art*, London, Macmillan.

Wolff, Janet (1989) 'The Invisible Flaneuse', in (ed.) Benjamin, 1989: 141–56.

Wolff, K. M. (1964) *The Sociology of Georg Simmel*, New York, Free Press.

Woodhouse, Tom (ed.) (1990) *Replenishing the Earth*, Bideford, Green Books.

Woods, Lebbeus (1995) 'Everyday War', in (ed.) Lang, 1995: 46–53.

Worthington, John (1996) 'Action Planning in the UK', *Urban Design*, conference proceedings, April: 12–14.

Young, James (1992) 'The German Counter-Monument', in Mitchell, 1992: 49–78.

Zukin, Sharon (1996) 'Space and Symbols in an Age of Decline', in (ed.) King, 1996: 43–59.

INDEX

•

Note: page numbers in italics refer to illustrations where these are separated from their textual reference.

abjection 50, 132, 161, 164, 173, 189, 223 (n6)
Adorno, Theodore 80
advertising 12, 54, 222 (n3)
advocacy 3, 95–7, 104–6, 108–12, 191
AIDS, art 16, 168–9, 172–6, 205
al-Khalil, Samir 67, 71
Albert Memorial 76
Alberti, Leon Battista 23, 33, 47
Alexander, Christopher 191
alienation 27, 62
allegory 69–73
Allen, Jerry 97
Ambrose, Peter 107, 124
Anaximander fragment 30, 213 (n22)
Anderson, N. 25
Andors, Rhoda 133
Andrews, Richard 94
anti-monuments 8, 80–3
Apgar, Garry 205
Appleyard, Donald 190, 192
architecture 43, 87–90, 112, 153
Arcosanti, Arizona 192
Armajani, Siah 202, 203
art: *see* public art
Art for Architecture (DoE) 95–6
The Art of Change 178–9
Art in a City (Willett) 91–3
art history 53–4, 55
Art Studios 169–72
Art Within Reach (Townsend) 93–4
artists: and architects 153; in community 207; complicity/criticism 147–8, 151; gender 53–4; in residency 127; socialist countries 86, 209 (n1)
Artists Agency 126–7, 169, 172
Arts Council 5, 88, 95, 96, 108–9, 111, 210 (n18)
asylum 32
Athens 30–1, 212 (n19)

Atherton, Kevin 8, 79, 77, 145, 219–20 (n22)
Australia, indigenous culture 181
Avalos, David 176

Baghdad, *Victory Arch* 71, 73
Balch, Clifton 25
Baratloo, Mojideh 25
Barcelona 205
Barthes, Roland 39, 42–3
Bartholdi, Frédéric-Auguste 71, 73
Battery Park City 119–24; conflict 225 (n30); critics 94, 122–4; *East Coast Memorial 82*, 83; feeding prohibitions 34, 35; homeless 106; playground use 236 (n9); railings 202; Winter Gardens 26
Batty, Dora 142
Baudelaire, Charles 17, 25, 34
Baudrillard, J. 186
Beardsley, John 75, 82, 97, 122
Becker, Carol 101
Beleschenko, Alexander 145
Benjamin, Andrew 47–8, 48
Benjamin, Walter 34, 63
Bentley, Ian 205
Berman, Marshall 13, 25, 32, 43
Beuys, Joseph 8, 183, 205
Bigelow, Kathryn 27
Bird, Jon 43, 67, 68, 106, 124
Birmingham, public art 115–17
Blackthorn Medical Centre, Kent 163
Blais, Jean-Charles 148
Bloch, Ernst 190
body, representations 30, 56, 61
Bookchin, Murray 194, 201, 237 (n19)
Border Art Workshop 176
Borenius, Tancred 76
Borofsky, Jonathan 5, 6
Boston 184–5, 191

Botero, Fernand 53, *119*
Bottle of Notes (Oldenburg and van
    Bruggen) 98–9, 104
boundaries 29, 31
bourgeois society 62, 145
Bradley, Laura 135, 203
Brancusi, Constantin 8, *12*
Brenson, Michael 167
de Bretteville, Sheila 178
Brighton, Andrew 85, 88–9, 94
British Rail 77
Broadgate development, London 53, 89,
    104, 110, 119
Brozgold, Lee 135, 146
Brunel, Isambard Kingdom 147
Bryson, Norman 42
Bucharest 23, *24*
Buck-Morss, Susan 63
Burgess, E. W. 35–6, 106, 118, 199
Burnham, Linda F. 169
Burton, Richard 110
Burton, Scott 8, 123, 203
Byatt, Lucy 160

Cable Street Mural 8, 71, *72*, 219 (n14)
Calder, Alexander 5
Calvino, Italo 42
capitalism 62, 200
Carde, Margaret 183
Cardiff Bay 112–14
Carr, Stephen 196
Cartesian dualism 45, 48, 56, 100–1,
    166–7, 207
Castells, Manuel 123, 177, 200
Ceaucescu, Nicolae 23, *24*
de Certeau, Michel 19, 22, 27, 36, 41–2
Cheere, John 60
Chicago School of Sociology 25, 34–6,
    236 (n8)
Chin, Mel 176, 182, 185
Cincinnati Gateway 74–5, 106
citizenship 201
city: alienation 27; as artefact 31; blood
    circulation metaphor 32; boundaries
    29; concept and form 19, 23;
    construction/destruction 43; exclu-
    sion/confinement 32–4, 213 (n27);
    future/past 28–9; gendered
    concepts/spaces 43; identity through
    art 117; images 25–9; legibility 193;
    liveableness 17, 202–3; popular
    culture 25–8; resistance 123; signifiers
    41; social reality 193, 194–9; social
    value 25–9; sustainability 192; views
    20–4; weaving metaphor 30–1, 213
    (n24); *see also* urban development;
    urban planning

city types: archaic 29–31, 212 (n13,
    n19), 213 (n21, n23); ideal 23;
    medieval 28–9, 46; modern 25,
    29–30, 189
civil society 29, 67, *68*
civitas 28, 87
Clinch, John 119, *120*
Coles, Peter 153
commodification 62, 75, 164
communications theory 37
community 178–9, 201; alternative 192,
    212 (n10); artists 207; as context 93;
    general public 99–100; involvement
    199; local perceptions 214 (n39);
    ownership of art 135; transport
    systems 137; urban planning 191,
    201
confinement, exclusion 32–4, 213
    (n27)
Constant, Christine 155, *158*
Contini, Anita 94
Coombes, Annie 63
Cooper, Robert 138
Coppinger, Siobhan 115
Cork, Richard 98–9, 153, 203
corporate fortress 119, 193
corporate greed, and social good 111
Covent Garden, London 107
craftwork 202–5
Creative Time 167–8
Cressey, P. G. 25
Cruikshank 95–6, 97
cultural industries 108
culture: affirmative 62–3; diversity
    176–9; domination 85–6; exclusion
    zone 51–4; heritage 74, 75, 88, 106,
    143; identity 58, 73–4; indigenous
    181; popular 14, 25–8, 143, 202
Czech Republic, public space 199

Dallas 114
Darke, Jo 60–1, 77
Davidoff, Paul 190–1, 200–1
Davidson, Gordon 188, 192, 208
Davis, Mike 17, 25, 155, *156*, *157*
death 151, 160–1, 220 (n23), 232 (n44,
    n45)
deconstruction 165–6
Delacroix, F. V. E. 70
Denny, Robyn 138
Denver, Catherine 143
Derkert, Siri 147
Descartes, René 45, 48, 56, 100–1,
    207
destruction 25, 43
Deutsche, Rosalyn 29, 90, 94, 97, 100,
    106, 122–4

DHSS 153
difference 57, 63, 176–9
Diggs, Peggy 15, 100, 102, 167–8, 169, 184, 207
digital imaging 179
Dine, Jim 49
disorder 57, 127, 190, 213 (n29)
Domestic Violence Milk Carton 167–8
domination, cultural 85–6
Dormer, Peter 15, 138
Doss, Erica 75
Douglas, Michael 138
Dring, Lilian 142
Drummond, Alan 144
Duncan, James 36, 43
Dunn, Peter 178–9

East Coast Memorial 82, 83
Eco, Umberto 89
ecological factors, barrage 113
ecological healing 182–6
Edward II, Swansea *130*
electronic technology 28, 179
Elsen, Albert 77
empowerment 43, 188; hospital staff 159; participation 103, 188; patient 162–3; performance art 100; urban dwellers 43, 188, 199–202; workers 76–7
Epstein, Jacob 91
exclusion 32–4, 48–54, 213 (n27)
exclusivity 124

Fanny Adams group 53
Fanon, Frantz 66–7
fantasy 62–3, 88
Featherstone, Mike 101, 109
Felshin, Nina 55, 100, 102
feminine 44, 56–7, 70
feminist cultural criticism 88
feminist writers 53–4
Feuer, Wendy 134, 135
Findhorn Community 192
Fisher, Mark 105
Fishman, Robert 25
Forester, John *37*
Forward (Mason) 77, 78, 115
Foster, Hal 102
Four Continents (French) 63–5
French, Daniel Chester 63–5
Freshman, Phil 81, 169
Frohnmayer, John 185
Fuller, Peter 97–8
Furuta, Hideo 118

Gablik, Suzi: connectedness 56, 188; ecological art 182; new genre public art 172; *The Reenchantment of Art* 100–1, 102, 164–5, 183, 207–8; on Serra 53, 89–90; simulation/reality 186–7
Gare, Aaren 201–2
gaze: masculine 44, 50, 54–5; medical 37–8, 161, 163, 232 (n43); planning 37–8; surveillance *21*, 197; tourist 74
gendering: artists 53–4; body heat 50; concept/space 1–2, 30, 43, 44, 45, 47, 50–1; difference 56–7; gaze 44, 50, 54–5; representation 47
Gentleman, David 138
gentrification 107–8
geography, women 53–4
George Washington sculpture 66
Gertz, Jochem 8, 80
Ghirardo, Diane 43, 100
di Giorgio, Francesco 41
Girardet, Herbert 28, 191
Glaser, Milton 135
Glasgow, European City of Culture 96
Goldsworthy, Andy 182
Gomila, Francis 97, 104, 222 (n1)
Goody, Joan 191
Gormley, Antony 5, 7, 87, 115, *116*, 224 (n24)
Gorr, Ted 174
Goya y Lucientes, Francisco José de 71
graffiti 134, *195*, 206
Gramsci, Antonio 66, 68, 123, 200
Gran Fury 173, 174, *175*
Greek city 29, 237 (n19); *see also* Athens
Green, Lynne 115, 117
Green, Oliver 134, 142
Greenberg, Clement 43
Greene, Lesley 147–8
Grey, Annabel 138, 143
grid planning 30–1, 42, 211 (n3)
Griffin, David Ray 167
Grimshaw, Nicholas 145
Griswald, Charles 81–2
Guilbaut, Serge 86
Guimard, Hector 133, 227 (n11)

Haacke, Hans 98
Haas, Richard 5, *10*
Habermas, Jurgen 37
Haha group 174, 176
Halbreich, Kathy 94
Hall, Peter 199
Hamilton, David 138, 143
Harris, Stacey Paleologos 91, 94
Harrison, Helen and Newton 182
Harvey, David 29–30, 101, 194, 200

Harvey, William 32, 48
Hasleden, Ron 115
Haussmann, George Eugène 23, 32, 37, 63, 213 (n26)
Hayden, Dolores 80, 101, 176, 177, 193, 198
health care, role of art 160–3; *see also* National Health Service
Heap of Birds, Edgar 102, 177, 179, 180, *181*
Heartney, Eleanor 165, 179, 182, 186
Heath, Jane 96
hegemony 66–7, 68, 123, 128–31, 220 (n4)
Henri, Adrian 91
heritage culture 74, 75, 88, 106, 143
Hernandez, Esther 173
Herrick, F. C. 143
Hewison, Robert 74
history and art 63–73, 81
Hock, Louis 176
Hogarth, William 151
Hoheisel, Horst 80
Holden, Charles 142
Holohan, Charles 151
Holt, Nancy 135
Holtzer, Jenny 8, 80
Holub, Renate 66–7
homelessness: Battery Park City 106; London Underground 144; Los Angeles 169; New York 122; as pollution 105; railway stations 145
hooks, bell 176, 177, 217–18 (n39)
Horace Greeley sculpture 65–6
Horne, David 74
House (Whiteread) 5, *9*, 101, 148, *149*, 206
Hughes, Graham 77
Husain, Saddam 71
Hutchinson, Fiona 153
Huxley, Paul 138
Huysmans, Joris Karl 63
hydroponics 174, 176

identity 58, 59–61, 73–4, 117, 218 (n2)
Illich, Ivan 33, 49, 151, 161, 169, 200, 208
individualism in art 15, 205
institutions 15, 85–6, 152–3, 209–10 (n13), 221 (n13)
integration, public art 203, 205
interiority 44, 47–54
internet 8, 209 (n10)
interventionist art 56, 205–8
investment art 110, 220 (n4)
Irigaray, Luce 39, 54–5, 57, 217 (n35)

Jackson, J. B. 25, 28–9, 41
Jackson, Peter 182
Jacob, Mary Jane 8, 100, 167, 176, 205
Jacobs, Allan 190, 192
Jacobs, Jane 34, 111, 133, 193, 197
Jacobs, Jane M. 181–2
Jagger, Charles Sergeant 69
Jaray, Tess 5, 8, *11*, 115, 203, 205
Jaudon, Valerie 135, *136*, 203
Jencks, Charles 87, 93–4
John, Saint 27
Johnson, Edward 134

Kaiser, Leland 160
Kaprow, Alan 102
Karavan, Dany 8
Kastner, Jeffrey 8
Kauffer, E. McKnight 142–3
Keep, P. 159
Kings Fund 160
Kluge, Alexander 123
Korza 95–6, 97
Kruger, Barbara 8, 55, 177–8

La Trobe-Bateman, Richard 203
Laboulaye, Edouard de 71, 73
Laclau, Ernesto 75
Lacy, Suzanne: *Crystal Quilt* 8, 100; intervention 205; new genre public art 55–6, 101–2, 164, 183, 184; performances 167, 200; public art framework 97; social healing 169
Laenen, Jean-Paul 147
Lang, Fritz 25, 27
Lang, Peter 17, 25
Le Corbusier 27
Leeson, Loraine 178–9
Lefebvre, Henri 39, 41, 44–6, 61, 122–3, 177, 198, 200
LeGates, Richard 190, 192, 199, 200, 201
legitimacy 67–9, 87–8
Leicester, Andrew 74–5, 135
Levitas, Ruth 101
liberty, allegory 70, 71, 73
Liberty, Statue of 71, 73–4, 219 (n16)
Lin, Maya 15, *81*, 81, 135
Lippard, Lucy 165, 167, 177
Liverpool 91–3
local authorities, public art 96
local people, public art 119, 124, 126–8, 226 (n34)
Loftman, Patrick 118
Lomax, Tom 115, 203
London: *Albert Memorial* 76; *Ash Wall* 205; Broadgate development 53, 89, 104, 110, 119; Coin Street 191;

Covent Garden 107; Docklands 105, 106, 107–8, 124–5, 178; Holly Street 191–2; *Nelson's Column* 76; Trafalgar Square *20*, 76; Villiers Street *21*
London Underground: 'Changing Stations' 136–42; Johnson's sans-serif typeface 134; modernity 142–3; nostalgia 143–4
Los Angeles 42, 43, 169, 177
Lowe, Nicholas 174
Lynch, Kevin 38, 193

McCarty, Marlene 173
McEwen, Indra Kagis 30, 46
McEwen, John 93
McInnes, Shona 153
Mackay, David 205
Mackie, Jack 146
McLaughlin, Corinne 188, 192, 208
McLean, Bruce 125
McMillan, Michael 174
Magnus, Dieter 205
Maine, John 203
Malpede, John 169
Manglano-Ovalle, Inigo 172
Manhattan *22*, *40*
Mapping the Futures 101
Mapping the Terrain 101–2
Marcuse, Herbert 58, 61, 62–3
marginalisation 31, 34, 49–50, 105
masculine 56–7; *see also* gaze, masculine
Mason, Biddy 177
Mason, Raymond 77, 78, 115
Massey, Doreen 101; gendered space 39, 45, 50, 51; masculine view 54; space/time 75–6; universals 88; visual sense 111
Mazeaud, Dominique 8, 182, 183–4, 235 (n29)
meaning, production 67
medical science 56
mega-cities 189
men: as artists 53; gaze 44, 50, 54–5; masculine principle 56–7; see also gendering
metro stations, public art 133–6, 147–8, 227 (n11); *see also* London Underground
Metropolis (Lang) 25, 27
Meyer, Richard 173, 174
Miles, Malcolm 119
Miners' Memorial, Frostburg 74
Miss, Mary *121*
Mistry, Dhruva 115, 117
Mitchell, Juliet 34
Mitchell, W. J. T. 73
Mitchell, William 61

modernist art 13, 86–7, 100–1, 165–7
modernity: city 25, 29–30, 189; exclusion/confinement 33; functionalism 133; nostalgia 142–4
Moffat, Donald 173
Molyneux, John 74
monument 73–6; anti-monuments 8, 58, 80–3; colonialism 218 (n1); cultural identity 58, 73–4; democratised 76–83; history/hegemony 63–73; national identity 58, 59–61
Monument Guide (Darke) 60–1, 73
Moore, Henry 5, 16, 89, 97
Moscow metro 133, 227 (n12)
Moses, Robert 25
mother–child image 54, 55, 57
Mouffe, Chantal 190, 205, 235 (n4)
Mozingo, Louise 50–1, 198
Mumford, Lewis 19, 28, 29, 212 (n13)
Munro, Nicholas 138
murals 8, 135
Myerscough, John 96, 108, 118

Nairne, Sandy 88
National Health Service: empowerment, patients/staff 159, 162–3; institutional art 152–3; national demonstration projects 154–60; patients' abjection 161; public art 150; visual surroundings/recovery 153, 158–60, 231 (n36)
native peoples, art 177, 179–82
naturalism 66, 70
Negt, Oskar 123
Nelson's Column 76
Nevin, Brendan 118
new genre public art 8, 55–6, 102–3, 164, 172, 183, 184
New York: City Hall Park sculpture 66; Custom House monuments 63–5; ecology/art 185; Greenacre Park 111, 196, *197*; homelessness 122; Paley Park 111, 194–6; subway 134–6; zoning regulations 193, *198*; *see also* Battery Park City; Manhattan
Ngo, Viet 185–6, 187
NHSE 163, 200
Nochlin, Linda 53
Noguchi, Isamu 86, *87*
nostalgia 106, 128, 142–4

Oldenburg, Claes 98–9
originality 15, 205
otherness 176; *see also* difference
Otterness, Tom 147, *148*
Owens, Craig 54

ownership: land/space 214–15 (n4);
    public art 135, 205
Ozymandias (Shelley) 67, *68*

Pantheon, Rome 67–8
Paolozzi, Eduardo 138, *139*
Papanek, Victor 208
Paris: Baudelaire 25; crime 28;
    Haussmann 23, 32, 63; sexualised
    215 (n15)
Paris Metro 133, 148, 227 (n11)
Park, Robert 35, 37
participation 97, 103, 166, 188
Partners for Livable Places 190
patients: abjection 161; empowerment
    162–3; recovery 153, 158–60, 231
    (n36)
patriarchy 44
Patten, David 115
Patterson, Ian 160
people of colour 176
Percent for Art 5, 104, 110–11, 135,
    146–7, 164
performance art 100, 108
perspective 23, 46
Petersfield, equestrian statue 60–1
Petherbridge, Deanna 87–8, 95, 115,
    137
Phaophanit, Vong 205
Philbin, Ann 173
Phillips, Patricia 14–15, 99–100, 102,
    164, 167–8, 184, 207
photography, by women 55
Picasso, Pablo 45, 61
Pick, Frank 134, 136, 142, 227 (n16,
    n19)
Piper, Adrian 176
place 74, 75; *see also* space
Planetree community health organisation
    161, 162–3
planning: *see* urban planning
Platforms Piece 77, 79, 80, 219–20
    (n22)
Plunz, Richard 17
polis 29, 237 (n19)
Pollock, Griselda 53, 55
pollution 105–6
poor, marginalised 31, 34
popular culture 14, 25–8, 143, 202
posters: AIDS 16, 173, *175*; Guerrilla
    Girls 52, 53; resistance 125, 148;
    transport 135–6, 142–3, 148, 228
    (n19)
postmodernism 17, 165
The Power of Place, Los Angeles 101,
    177–8
psychiatric patients, art studios 170–1

public art 1, 5–12; activism 84;
    advocacy 3, 95–7, 104–6, 108–12,
    191; criticism 3, 97–102; funding 5,
    96, 115–28; health 150, 151–2,
    154–60, 161, 162–3, 231 (n36);
    histories 101–2; integration 203, 205;
    interventionist 56, 205–8; as invest-
    ment 110, 220 (n4); literature 3–4,
    91–4; local authorities 96; and local
    people 119, 124, 126–8, 226 (n34);
    monument 61; new genre 8, 55–6,
    102–3, 164, 183; participation 97,
    103, 166, 188; problematised 85–90;
    public reception 12–15, 92, 238
    (n35); social comment 147; as social
    good 16, 111, 132; in streets 59;
    tourism 113; transport systems
    132–49; urban decay 17; *see also*
    institutions
public realm 100, 164, 207
Public School, USA 135, 146
public space 1–2, *40*, 184, 193–202
public transport 132–3, 136–43, 144–9
purification, abjection 50, 223 (n6)
Pym, William 128

railways 145, 147; *see also* metro stations
Randall-Page, Peter 182, *183*
Raven, Arlene 85, 100, 165, 167
reality/simulation 186–7
Reason, David 113, 225 (n25)
reason, interiority 47–54
reclamation 56–7
Red House 126, 128, *129*
The Reenchantment of Art (Gablik) 100–1,
    102, 165, 183, 207–8
regeneration 112–13
Reiter, Wellington 138, 142, 147
representation: gendered 47; problem
    41–3; space 39, 44–7, 56, 59
resistance 123; accepting difference 176;
    city 123; digital imaging 179;
    monument democratised 76–83;
    posters 125, 148; strategies 54–7;
    through vandalism 205
Rigler, Malcolm 163
Rilke, Rainer Maria 25, 211 (n7)
Ringgold, Faith 146
Rittner, Luke 108
ritual 184
road-building protests 148
Roberts, Marion Salter 96
Roberts, William 143
Rodgers, Peter 108
Rodin, Auguste 76
Rogers, Alan 143
Rogers, Richard 105

Rollins, Tim 172
Rome, Pantheon 67–8
Rose, Barbara 53–4
Rose, Gillian 50, 56
Rosler, Martha 101, 102, 107, 123, 169
Ross, Nick 159
Roth, Moira 167
Rowland, Jon 201
ruin value theory 67
Russia, public sculpture 59
Rykwert, Joseph 29–30
Rysbrack, Michael 61

Salter, Miffa 193, 198
Sandle, Michael 68
Sassen, Saskia 106, 110, 114, 118
Savage, Mike 36, 37
Savitch, H. V. 107, 119
Scott, Gilbert 152
Scott, Giles Gilbert 76
Scott, Joyce 8, 13
sculpture 8, 16, 59, 112
Seattle 146–7, 179–80, 195
Selwood, Sara 91, 96–7, 105, 110, 115, 118, 182
Sennett, Richard: circulation of city 32; disorder 57, 127, 190; exclusion 33, 192; Flesh and Stone 30, 50; graffiti 134; interiority 44; legitimacy 68; medieval city 28, 31, 191; public transport 133; urban diversity 17, 18
sense impressions 47–8, 216 (n24)
Serra, Richard 5, 53, 87, 89–90, 205, 208
Shalev-Gertz, Esther 8, 80
Shamash, Diane 180
Shelley, Percy Bysshe 67, 68
Sherman, Cindy 55
Shields, Rob 25, 39, 41
Short, John Rennie 38, 194
Sibley, David: car adverts 222 (n3); Cardiff Bay 113; city boundaries 29, 31; gaze 54; pollution/purity 105–6, 223 (n6); purification/abjection 49–50; zoning 35
Sierhuis, Jan 148
signifiers 39, 41, 215 (n11)
Simmel, Georg 34
simulations, and reality 186–7
Sisco, Elizabeth 176
slum-clearance 27–8
Smith, Adam 32
Smyth, Ned 121
social comment 147–8
social good 132, 189
social healing 169–76
social life, urban space 194–9

sociation 36, 112, 144–9, 172
Soleri, Paolo 192
Sonfist, Alan 185
Sontag, Susan 161
Sorkin, Michael 191, 193–4
space: conceptions 46; gendered 30, 43, 44, 45, 47, 50–1; ideological 122–3; internet 8, 209 (n10); ownership 214–15 (n4); production 44–5; public/corporate 1–2, 40, 184, 193–202; representational 39, 59; representations 39, 44–7, 56, 59; time 44–5, 75–6
Speer, Albert 67
splitting 33–4, 166–7
Spretnak, Charlene 165, 166, 187
St Louis, rail system 147
stained glass 152–3
statues 60–1, 71, 73–4, 77, 81; see also sculpture
Stockholm Metro 147
Stout, Frederic 190, 192, 199, 200, 201
Strange Days 191
street furniture 203
street life 194–9
street plans 41–2
stress, visual surroundings 158–60
suburbia 25
Subversive Imagination 101
subways: see metro stations
Sunderland 125–8
surveillance 21, 197
sustainability 105, 188, 192–3, 207–8
Swindon, public art 118–19
Sztaray, Susan 178

tapestry 153
taste 16, 92, 97, 111, 152
Tebby, Susan 153
Tele-Vecindario project 172
Thamesdown Borough Council 118–19
tiling 138, 152
Tilted Arc 89, 90, 165
time, space 44–5, 75–6
Tipping, Chris 138, 143, 144
tourism 74, 113, 143, 179, 181–2
Townsend, Peter 91, 93–4
toxic waste 185
Trafalgar Square 20, 76

Ukeles, Mierle Laderman 8, 102, 182, 183, 184, 200
Ulrich, Roger 159
underclass 27, 134
urban crisis 17, 194, 208
Urban Design Group 192–3, 200–1

urban development 105–6; alternative discourse 190–3; case studies 115–28; future 189; hegemony 128–31; local people 107–8; regeneration 112–13
urban dwellers 43, 188, 199–202
urban ethnography 34–5
urban planning: action planning 201; Baroque to Enlightenment 23; community-centred 191, 201; grid system 30–1, 42, 211 (n3); marxism 200; public art 188; sociology 36; zoning 31–8
Urban Renaissance (Arts Council) 108–9, 113
Urry, John 74
USA: art and participation 97; health and art 152, 153, 154; public art 95–6; public transport 144–9; *see also* individual cities

value structures 62, 67
van Bruggen, Coosje 98–9
Vance, Carol 168
vandalism 128, 130, 205
Venice 31
victory, allegory 71, 73
Victory Arch, Baghdad 71, 73
Villiers Street, London *21*
Vilmouth, Jean-Luc 145
violence 61, 169, 172, 208, 217 (n29)
visual art, privileged 111
Visual Dallas 114
visual surroundings, health 153, 158–60
Vorticism 143

Wallis, Brian 100, 169
Walters, Ian 66
war, idealised in art 68–9
war memorials 61
Ward, Dick 153
Ward, John 66
Warde, Alan 36, 37
Warner, Marina 70, 73–4
Washington metro 133–4
waste recycling 184

Wates, Nick 201
Watkinson, Cate 155, *156*, *157*
Watney, Simon 173
Weekes, John 153
Weems, Carrie Mae 177
Wehn, James 179
Weinstein, Jeff 169, 174
White, John 47
Whiteread, Rachel 5, 9, 101, 148, *149*, *206*
Whyte, W. H. 14, 51, 94, 111, 114, 189; Street Life Project 194–7, 199
Widgery, David 124–5
Wigley, Mark 27, 29, 33, 38, 42
Wilbourne, Colin 8
Willett, John: allegory 70; Arts Council 111; Liverpool 91–3; monument 61; public art 85; St John's transformed city 27; taste 16, 95
Wilson, Elizabeth 23, 31, 38, 51
Wilson, Larkin 159
Winkel, Gary 151
Wirth, Louis 36–7
Withymoor Village Surgery 163
Wodiczko, Krzysztof 8, 80, 81, 169
Wolff, Janet 51
women: art history 53–4; artists 100, 168; feminine principle 44, 56–7, 70; marginalised 49–50; medical gaze 218 (n42); representations 217 (n29); urban space 50–1; *see also* gendering
Wood, F. Derwent 69
Woodin, Mary 143
Woods, Lebbeus 25, 43

Young, Gordon 8, 203, *204*
Young, James 80

Zadkine, Ossip 92
zoning 31–4; Cartesian 45; as classification 38; concentric 35–6, 106, 118, 199; cultural exclusion 51–4; New York 193, *198*; social 105–6; by usage 34, 193
Zukin, Sharon 38, 59, 67, 110, 117